QUADRIVIUM

Published by Walker Publishing Company, Inc., New York
A Division of Bloomsbury Publishing

All papers used by Walker & Company are natural, recyclable products
made from wood grown in well-managed forests. The manufacturing processes
conform to the environmental regulations of the country of origin.

LIBRARY OF CONGRESS CATALOGING-IN-PUBLICATION
DATA IS AVAILABLE.

ISBN: 978-0-8027-7813-0

Visit Walker & Company's Web site at www.walkerbooks.com

Sacred Number was originally published by Walker & Company in 2005
Sacred Geometry was originally published by Walker & Company in 2001
Platonic & Archimedean Solids was originally published by Walker & Company in 2002
Harmonograph was originally published by Walker & Company in 2003
The Elements of Music was originally published by Walker & Company in 2008
A Little Book of Coincidence was originally published by Walker & Company in 2002

5 7 9 10 8 6 4

Designed and typeset by Wooden Books Ltd, Glastonbury, UK
Printed in China by South China Printing Co., Dongguan, Guangdong

QUADRIVIUM

THE FOUR CLASSICAL LIBERAL ARTS OF
NUMBER, GEOMETRY, MUSIC, & COSMOLOGY

Walker & Company
New York

CONTENTS

Editor's Preface

The volume you are holding is a rare treasure chest of old, once secret, and always useful things. It is also immortal—it will never go out of date. Thirdly, it is universal—a magical passport between cultures, sacred and scientific, provincial and foreign, ancient and modern.

Six books from the Wooden Books series have been combined to produce *Quadrivium*, plus 32 new pages for good measure. Geometry and music are each covered in two books, hence the six. We have tried to minimize repetition due to overlap in the original books but a few instances remain here and there. Also, since this is a transatlantic edition, we have opted for a mixture of English and American spelling and punctuation. If this upsets some readers, we apologize.

A volume of this scope involves bringing together many people. Thanks to Sally Pucill, Richard Henry, Adam Tetlow, John Michell, John Neal, Dr Paul Marchant, Robin Heath, David Wade, Dr Khaled Azzam, Malcolm Stewart, Polly Napper, Geoff Stray, Dr Moff Betts, Prof Scott Olsen, Richard Heath, Matt Tweed, Mark Mills, Prof Robert Temple, Stephen Parsons, Nathan Williams, Charlie Dancey, and Tracey Robinson for their help and contributions.

Thanks to the additional Wooden Books series editors, George Gibson of Walker & Bloomsbury in New York, and Daud Sutton in Cairo. Thanks finally to Prof Keith Critchlow for the foreword, and to the authors themselves, Miranda Lundy, Daud Sutton, my grandfather Anthony Ashton, and Dr Jason Martineau.

John Martineau

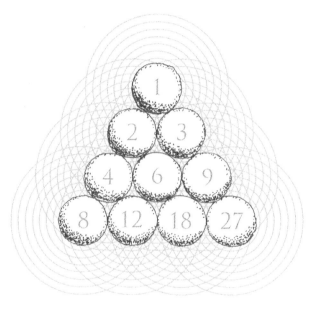

The TETRATKYS of the Pythagoreans augmented by the LAMBDA
of the Timaeus. Plato kept three numbers back, only revealing
seven: 1, 2, 3, 4, 8, 9, and 27, as related to the planets.
Pebbles or khalix were the Greek norm for mathematics.

FOREWORD

The Quadrivium was first formulated and taught by Pythagoras as the Tetraktys around 500 BC, in a community where all were equal, materially and morally, and where women had equal status to men. It was the first European schooling structure that honed education down to seven essential subjects, later known as the seven liberal arts.

The word 'education' comes from the Latin *educere* meaning 'to lead out', pointing to the central doctrine that Socrates, under Plato's pen, elucidated so clearly—Knowledge is an inherent and intrinsic part of our soul structure. The *Trivium* of language is structured on the cardinal and objective values of Truth, Beauty and Goodness. Its three subjects are *Grammar,* that ensures the good structure of language, *Logic* for finding truth, and *Rhetoric* for the beautiful use of language in expressing truth. The *Quadrivium* arises out of the most revered of all subjects available to the human mind—Number. The first of these disciplines we call *Arithmetic*, the second is *Geometry* or the order of space as Number in Space, the third is *Harmony* which for Plato meant Number in Time, and the fourth is *Astronomy*, or Number in Space and Time. All these studies offer a safe and reliable ladder to reach the simultaneous values of the True, the Good, and the Beautiful. This in turn leads to the essential harmonious value of Wholeness.

Our soul, which is proven to be immortal by Socrates in the *Phaedo*, comes from a position of complete knowledge prior to being born into the body. Re-membering—the point of education—literally means putting the separate members back into a wholeness. The goal of studying these subjects was to climb back (up) to Unity through simplification based on the understanding gained by engaging in each area of the Quadrivium. The goal lay in finding their *source* (traditionally this was the sole purpose of the search for knowledge).

In his discussions on the ideals of education, Socrates reveals his model of the continuity of consciousness. This was as a 'line' drawn vertically reaching from the beginnings of conscious knowledge in Estimations right up to the climax of consciousness as *Noesis,* which is Unified Understanding. Beyond this is the indescribable and ineffable. There are, significantly, four stages (another quadrivium or *Tetratkys*) given by Socrates' division of the 'Ontological line'. The first and fundamental division is between the Intelligible and Sensory world, between Mind and Matter. Next, each of these is divided, Estimations from Opinions. In the sensory world even correct opinions are still based on sensory experience, while above the first divided line, in the Intelligible world of the Mind we find ourselves in the 'truth-bearing' realm of the Quadrivium, true and objective knowledge. The final topmost division of the intelligible is *Nous* or Pure Knowledge itself where the knower, the known, and the knowing become One. This is the goal and source of all knowledge. Thus, time and wisdom tested, the Quadrivium offers the sincere seeker the opportunity to regain their own inner understanding of the integral nature of the universe, with themselves as an inseparable part.

Turning now to the 'four ways' in more detail, *Arithmetic* has three levels: the materially Numbered; Mathematicians' Number (indefinite); and Ideal or Archetypal Number (complete at 10).

Geometry unfolds in four stages: the non-dimensional point; which moves to become a line; which in turn moves to become a plane; finally achieving solidity as the tetrahedron. *Harmony* (the nature of the Soul) displays four musical 'scales', the pentatonic, diatonic, chromatic, and shruti. Finally we come to the *Cosmos*, a word originated by Pythagoras, meaning 'order' and 'adornment'. The Pythagoreans viewed the visible heavens as an 'adornment' of pure principles, the number of visible planets relating to the principles of proportional harmony. The study of the perfection of the heavens was a way of perfecting the movements of one's own soul.

Students of the Quadrivium have included: Cassiodonus, Philolaus, Timaeus, Archytus, Plato, Aristotle, Eudemus, Euclid, Cicero, Philo the Jew, Nichomachus, St. Clement of Alexandria, St. Origen, Plotinus, Iamblichus, Macrobius, Capella (his is the most entertaining version available), Dionysius the Areopagite, Bede, Alcuin, Al-Khwarizmi, Al-Kindi, Eriugena, Gerbert d'Aurillac, the Brethren of Purity, Fulbert, Ibn Sina (Avicenna), Hugo of St. Victor, Bernardus Silvestris, Bernard of Clairvaux, Hildegard of Bingen, Alanus ab Insulis, Joachim of Fiore, Ibn Arabi, Grosseteste (the great English scientist), Roger Bacon, Thomas Aquinas, Dante, and Kepler.

We finish with two quotes. The first is from the Pythagoreans (via the *Golden Verses*): "AND THOU SHALT KNOW THAT LAW ... ESTABLISHED THE INNER NATURE OF ALL THINGS ALIKE"; The second is from Iamblichus: "NOT FOR YOUR SAKE WAS THE WORLD (COSMOS) GENERATED—BUT YOU WERE BORN FOR ITS SAKE."

Keith Critchlow

BOOK I

Gregor Reisch's sixteenth-century engraving showing Pythagoras using
a medieval counting board to form the numbers 1,241 and 82 (*right*)
while Boethius calculates using the Indian numerals we are familiar
with today (*left*). In the center is Arithmetic, with the two geometric
progressions 1,2,4,8, and 1,3,9,27 appearing on her dress.

SACRED
NUMBER

The Secret Qualities of Quantities

Miranda Lundy

with additional material by Adam Tetlow & Richard Henry

INTRODUCTION

What is number? How do we distinguish the one from the many, or, for that matter, the two from the three? A crow, disturbed by four men with guns going to hide under its tree, will fly away and carefully count them home again from a distance, one by one, tired and hungry, before returning safely to its nest. But five? Crows lose count at five.

We all know certain things about certain numbers: six circles fit around one, there are seven notes in a scale, we count in tens, three legs make a stool, five petals form a flower. Some of these elementary discoveries are actually the first universal truths we ever come across, so simple we forget about them. Children on distant planets are probably having the same experiences of these elementary quanta.

The science and study of number is one of the oldest on Earth, its origins lost in the mists of time. Early cultures wrote numbers in pottery markings, weaving patterns, notched bones, knots, stone monuments, and the numbers of their gods. Later systems integrated the mysteries under the magical medieval Quadrivium of arithmetic, geometry, music, and astronomy—the four liberal arts required for a true understanding of the qualities of number.

All science has its origin in magic, and in the ancient schools no magician was unschooled in the power of number. These days the lore of sacred number has been usurped by a tide of quantitative numbers, not covered in these pages. BOOK I of *Quadrivium* is a beginner's guide to mystic arithmology, a small attempt to unveil some of the many secret and essential qualities of number contained within Unity.

THE MONAD

one unity

Unity. The One. God. The Great Spirit. Mirror of wonders. The still eternity. Permanence. There are countless names for it.

According to one perspective, one cannot actually speak of the One, because to speak of it is to make an object of it, implying separation from it, so misrepresenting the essence of Oneness from the start, a mysterious conundrum.

The One is the limit of all, first before the beginning and last after the end, *alpha* and *omega*, the mold that shapes all things and the one thing shaped by all molds, the origin from which the universe emerges, the universe itself, and the center to which it returns. It is point, seed, and destination.

One is echoed in all things and treats all equally. Its stability among numbers is unique, one remaining one when multiplied or divided by itself, and one of anything is uniquely that one thing.

All things are immersed in the shoreless ocean of Unity. The quality of oneness permeates everything, and while there is nothing without it, there is also no thing within it, as even a communication or idea requires parts in relationship. Like light from the Sun or gentle rain, the One is traditionally perceived as unconditional in its love, yet its majesty and mystery remain veiled, and beyond apprehension, for the One can only be understood by itself. As such, One is always alone, all one, and no thing can exist to describe it.

One is simultaneously circle, center, and the purest tone.

DUALITY

opposites

There are two sides to every coin, and the other side is where the Dyad lives. Two is the otherworldly shadow, opposite, polarized, and objectified. It is there, other, that not this, and essential as a basis for comparison, the method by which our minds know things. There are countless names for the divine pair.

To the Pythagoreans, two was the first sexed number, even and female. To develop their appreciation of twoness, they contemplated pairs of pure opposites, such as limited-unlimited, odd-even, one-many, right-left, male-female, resting-moving, and straight-curved. We might also think of the in-and-out of our breathing, positive and negative charges in electromagnetism, and real and imaginary numbers.

The dyad appears in music as the ratio two to one, as we experience a similar tone an octave higher or lower, at twice or half the pitch. In geometry it is a line, two points, or two circles.

Linguistically when speaking of both parts of something working as one we use the bi- prefix, as in bicycle or binary, but when the divisive quality of two is invoked, words begin with the prefix di-, thus discord or diversion. The distinction between self and not-self is one of the first and last we generally make.

Modern philosophers, when they consider twoness, can get little further than the ancients. All experience a left and a right, front and back, and up and down through two eyes and two ears. Men and women alike live under a Sun and a Moon, only occasionally remembering how miraculously balanced they seem, the same size in the sky, one shining by day, the other by night.

THREE
is a crowd

Male in some cultures, female in others, three, like a tree, bridges heaven and earth. The Triad relates opposites as their comixture, solution, or mediator. It is the synthesis or return to unity after the division of two and is also traditionally the first odd number.

The third leg of a stool gives it balance, the third strand of a braid makes a plait (knots can only be tied in three-dimensional space). Stories, fairy tales, and spiritual traditions abound with portentous threes, juggling past, present, and future with the knower, knowing, and the known. As birth, life, and death, the triad appears throughout nature, in principle and form. The triangle, trinity's most simple and structural device, is the first stable polygon, defining the first surface.

In music the ratios 3:2 and 3:1 define the intervals of the fifth and its octave, the most beautiful harmonies other than the octave itself, and the key to all ancient tunings. Three is the first triangular number.

The *vesica piscis* formed by two overlapping circles (*opposite top left*) immediately invokes triangles. An equilateral triangle in a circle defines the octave as its circumcircle is twice the diameter of is incircle (*below left*), and the area of the ring here is three times that of the incircle. Below center we see Archimedes' (287–212 BC) favourite discovery—the volumes of the cone, sphere, and drum are in the ratios one to two to three.

QUATERNITY
two pairs

Beyond three we enter the realm of manifestation. Four is the first born thing, the first product of procreation, two twos. The Tetrad is thus the first square number (other than one), and a symbol of the Earth and the natural world.

Four is the basis of three-dimensional space. The simple solid known as the tetrahedron, or 'four facer', is made of four triangles, or four points or spheres and is as fundamental to the structure of three-dimensional space as the triangle is to the plane.

Four is often associated with the material modes of manifestation, Fire, Air, Earth, and Water, and a square around a circle defines a heavenly ring whose area is equal to the enclosed circle (*opposite top right*). The solstices and equinoxes quarter the year, horses walk on four legs, and other earthly fours abound.

Four as static square is echoed by the dynamic cross. The interplay of cross and square is encoded within the traditional rite of orientation for a new building, where the sunrise and sunset shadows from a central pillar give the symbolic east-west axis. The principle of quadrature is universal, appearing in ancient Chinese texts and the writings of Vitruvius. It survives today in the term *quarters*, referring to the districts of a city.

All everyday matter is appropriately made of just four particles: protons, neutrons, electrons, and electron neutrinos.

Four appears in music as the third overtone, $4:1$, which is two octaves, and also as the ratio $4:3$, known as the fourth, which is the complement of the fifth inside the octave.

PHIVE

life itself

The quality of five is magical. Children instinctively draw fivefold stars, and we all sense its phizzy, energetic quality.

Five marries male and female—as two and three in some cultures, or three and two in others—and so is the universal number of reproduction and biological life. A Fibonacci number, it is also the number of water, every molecule of which is a corner of a pentagon. Water itself is an amazing liquid crystal lattice of flexing icosahedra, these being one of the five Platonic solids (*below, second from right*), five triangles meeting at each point. As such, water shows its quality as being that of flow, dynamism, and life. Dry things are either dead or they are awaiting water.

Fives are found in apples, flowers, hands, and feet. Our nearest planet, Venus, goddess of love and beauty, draws a lovely fivefold pattern about Earth as she whirls around the Sun (*opposite top left*).

Our most universal scale, the pentatonic, is made of five tones (the black keys on a piano), grouped into two and three. The Renaissance demand for intervals involving the number five, like the major third, which uses the ratio 5:4, produced the modern scale.

Five is the diagonal of a three-by-four rectangle. However, unlike threes and fours, fives disdain the plane, waiting for the third dimension to fit together to produce the fifth element.

ALL THINGS SIXY
the hex

The Hexad, like its graceful herald the snowflake, brings perfection, structure, and order. The marriage by multiplication of two and three, even and odd, six is also the number of creation, with a cosmos made in six days a common theme in scripture.

The whole numbers that divide other numbers are known as their *factors*, and most numbers have factors that sum to less than themselves, and so are known as *deficient*. Six, beautifully, is the sum and product of the first three numbers, and its factors are also just one, two, and three, these summing to six and so making it the first *perfect number*.

The radius of a circle can be swung through its circumference in exactly six identical arcs to inscribe a regular hexagon, and six circles perfectly fit around one. After the triangle and square the hexagon is the final regular polygon that can tile perfectly with identical copies of itself to fill the plane.

The three dimensions make for six directions: forward, backward, left, right, up, and down, and these are embodied in the six faces of a cube, the six corners of an octahedron and the six edges of a tetrahedron. Six occurs widely in crystalline structures such as snowflakes, quartz, and graphite, and hexagons of carbon atoms form the basis of organic chemistry. Just add water.

The well-known Pythagorean 3-4-5 triangle has an area and a semiperimeter of six. Six is also the pentatonic octave in music.

Insects creep and crawl on six legs, and the honey bee arranges its dry, waxy secretions into an instinctive hexagonal honeycomb.

THE HEPTAD
seven sisters

Seven is the Virgin, standing quite alone and having little to do with any of the other simple numbers. In music a scale of seven tones emerges as naturally as its sister five-tone scale. These are the white keys on the piano, producing the seven modes of antiquity, a universal pattern. Like all numbers, seven embodies the number preceeding it; spatially it functions as the spiritual center of six, as six directions emanate from a point in space, and six working circles surround a seventh restful one in a plane.

The Moon's phases are widely counted in four sevens with a mysterious moonless night or two completing its true cycle.

Our eyes perceive three primary colours of light—red, green and blue—which combine to produce four more—yellow, cyan, magenta and white. According to the ancient Indians, a vertical rainbow of seven subtle energy centers, or 'chakras', runs up our bodies. Today we understand these as the seven endocrine glands.

The seven planets of antiquity, arranged in order of their apparent speed (*opposite, upper center*), make amazing connections with metals (*opposite, upper left*) and the days of the week (*opposite, upper right*): Moon-☽-silver-Monday, Mercury-☿-quicksilver-Wednesday, Venus-♀-copper-Friday, Sun-☉-gold-Sunday, Mars-♂-iron-Tuesday, Jupiter-♃-tin-Thursday, and Saturn-♄-lead-Saturday (*see too page 305*).

There are seven frieze symmetries, seven groups of crystal structures, and seven coils in the traditional labyrinth (*all shown opposite*).

25

EIGHT
a pair of squares

Eight is two times two times two, and as such is the first cubic number after one. As the number of vertices of a cube or faces of its dual, the octahedron, eight is complete. At the molecular level this is displayed by atoms, which long to have a full octave of eight electrons in their outermost shell. A sulphur atom has six electrons in its outermost shell, so eight atoms get together to share electrons, forming an octagonal sulphur ring. Eight is also the next Fibonacci number after five.

Within architecture the octagon often signifies the transition between Heaven and Earth, as a bridge between the square and the circle. A spherical dome often surmounts a cubic structure by way of a beautiful octagonal vault.

Eight is particularly revered in the religion and mythology of the orient. The ancient Chinese oracle, the *I-Ching*, is based on combinations of eight *trigrams*, each the result of a twofold choice between a broken (yin) or an unbroken (yang) line, made three times. Depicted opposite is the 'Former Heaven Sequence', said to represent the ideal pattern of transformations in the cosmos. Note how each trigram is the complement of its opposite. Echoing these ancient Chinese pairs, the modern world is full of computers which think in units called 'bytes', each made of eight binary 'bits'.

In a seven-tone scale the eighth note is the octave, twice the pitch of the first note, and so signals the movement to a new level. This may be why, in religious symbolism, the eighth step is often associated with spiritual evolution or salvation.

27

THE ENNEAD

three threes

Nine is the triad of triads, the first odd square number, and with it something extraordinary occurs, for the first nine numbers can be arranged in a magic square where every line of three in any direction has the same total (*opposite center*). This ancient number pattern was first spotted four millennia ago on the shell of a divine turtle emerging from the river Lo in China.

Three times three is one more than two times two times two, and the ratio between nine and eight defines the crucial whole tone in music, the 9:8 seed from which the scale emerges, as the difference between the two most simple harmonies in the octave, the fifth 3:2, and the fourth 4:3.

There are nine regular three-dimensional shapes: the five Platonic solids and the four stellar Kepler-Poinsot polyhedra (*see Book III*).

Nine appears in our bodies as the cross-section of the tentacle-like cilia, which move things around our surfaces, and the bundles of microtubes in centrioles, essential for cell division (*below*).

Nine is the celestial number of order, and many ancient traditions speak of nine worlds, spheres, or levels of reality. Cats know. They have nine lives, dress to the nines whenever possible, and seem to spend most of their time on cloud nine, wherever that is.

TEN
fingers and thumbs

The fact that humans have eight fingers plus two thumbs must have worked in ten's favour, as cultures as various as the Incas, the Indians, the Berbers, the Hittites, and the Minoans all adopted it as the base for their counting systems. Today we all use base ten. Ten is the child of five and two, and unsurprisingly the word *ten* derives from the Indo-European *dekm*, meaning 'two hands'.

Ten is particularly formed as the sum of the first four numbers, so $1+2+3+4=10$, a fact of profound significance to the Pythagoreans who immortalized it in the figure of the Tetraktys (*black dots, opposite center*) and called it Universe, Heaven, and Eternity. As well as being the fourth triangular number, ten is also the third tetrahedral number (*lower, opposite right*), a fact that lends it great importance as a simultaneous building number of both two- and three-dimensional triangular form.

Ten is formed from two pentagons and ten life-invoking pentagons sit perfectly around a decagon, and DNA, appropriately as the key to the reproduction of life, has ten steps for each turn of its double helix, so appears in cross-section as a tenfold rosette (*opposite top left*).

There are ten *Sephiroth* in the Jewish Kabbalah's Tree of Life (*lower, opposite left*) and tenfold symmetry was often used in Gothic architecture (*opposite top right*).

Plato believed that the decad contained all numbers, and for most of us today it does, as we can express just about any number we care to think about in terms of just ten simple symbols.

ELEVENSES

measure and the Moon

Eleven is a mysterious underworldly number—in German it goes by the appropriate name of *Elf.* Eleven is important as the first number that allows us to begin to comprehend the measure of a circle. This is because, for practical purposes, a circle measuring seven across will measure eleven halfway around (*opposite top left*).

This relationship between eleven and seven was considered so profound by the ancient Egyptians that they used it as the basis for the design of the Great Pyramid. A circle drawn around the elevation of the Great Pyramid has the same perimeter as that of its square base. The intended seven-elevenfold conversion between square and curve is demonstrated by numerous surveys.

The ancients were obsessed with measures, and the number eleven is central in their metrological scheme. Shown opposite is the extraordinary fact that the size of the Moon relates to the size of the Earth as does three to eleven. What this means is that if we draw down the Moon to the Earth, as shown, then a heavenly circle through the Moon will have a circumference equal to the perimeter of a square around the Earth. This is called 'squaring the circle'. Quite how the old druids worked this out we may never know, but they clearly did, for the Moon and the Earth are best measured in miles, as shown. A double rainbow also magically squares the circle (*see page 78*).

Eleven, seven, and three are all Lucas numbers, sisters of the Fibonacci numbers, new numbers forming from the sum of the previous two numbers. The Fibonacci sequence begins 1, 1, 2, 3, 5, 8, whereas the Lucas sequence begins 2, 1, 3, 4, 7, 11.

MOON DIAMETER
= 3 × 1×2×3×4×5×6
= 3 × 8×9×10 MILES

EARTH DIAMETER
= 1×2×3×4×5×6 × 11
= 8×9×10 × 11 MILES

MOON ✴ EARTH RADII
= 1×2×3×4×5×6×7
= 7×8×9×10 MILES

AREA OF HEAVENLY CIRCLE
= 2 × 1×2×3×4×5×6×7×8×9×10×11
SQUARE MILES

THE TWELVE
heaven and earth

Twelve is the first *abundant number*, with factors one, two, three, four, and six, summing to more than itself. Twelve points on a circle can join to form four triangles, three squares, or two hexagons (*opposite center*). As the product of three and four, twelve is also sometimes associated with their sum, seven.

Twelve enjoys the third dimension and is the number of edges of both the cube and the octahedron. The icosahedron has twelve vertices, and its dual, the dodecahedron (literally 'twelve facer') has twelve faces of regular pentagons. Twelve spheres fit perfectly around one to define a cuboctahedron. We will meet these polyhedra later.

In a seven-note scale, notes increase as a pattern of five tones and two halftones. In modern tuning the five tones are divided to create a scale of twelve identical halftones, the well-tempered twelve tone scale we all hear every day.

Curiously, the next most simple Pythagorean triangle, after the three-four-five, has sides of five, twelve, and thirteen units.

Twelve is often found arranged around a central solar hero, and there are many twelve-tribe nations. In ancient China, Egypt, and Greece, cities were often divided into twelve districts, and, of course, there are usually twelve full moons in a year.

The material universe is today understood as being made of three generations of four fundamental particles, twelve in all.

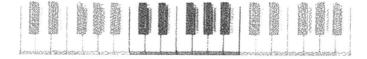

COVENS AND SCORES
into higher numbers

Thirteen, the coven, beloved of the ancient Maya, and central to the structure of a deck of cards, is a Fibonacci number expressed in the motions of Venus, for whom thirteen years is eight of our own, and lest you think it unlucky, remember the teacher of twelve disciples is the thirteenth member of the gang, as the thirteenth tone in the chromatic scale completes the octave.

Fourteen, as twice seven, and fifteen, as three fives, each have unique qualities but begin to demonstrate how non-prime higher numbers tend to be perceived in terms of their factors.

Sixteen is $2 \times 2 \times 2 \times 2$, the square of four (itself a square).

Seventeen keeps many secrets. Both Japanese haiku and Greek hexameter consist of seventeen syllables, and Islamic mystics often refer to it as particularly beautiful.

Eighteen, as twice nine and thrice six, and nineteen, a prime number, both have strong connections to the Moon (*see page 42*).

Twenty, a score, the sum of fingers and toes, is a base in many cultures. Finger-counting, as in the example shown (*opposite*), was widespread in medieval European markets. In French eighty is still *quatre-vingt* (four twenties) and the ancient Maya used a sophisticated base-20 system (*glyphs for 1–19 shown below*).

There is not enough space here to cover every number in detail, but interesting facets of higher numbers appear in the glossary of numbers at the back of the book (*see pages 364–366*).

· ·· ··· ···· ▬ ▬ ▬ ▬ ▬ ═ ═ ═ ═ ═ ▬ ▬ ▬ ▬ ▬

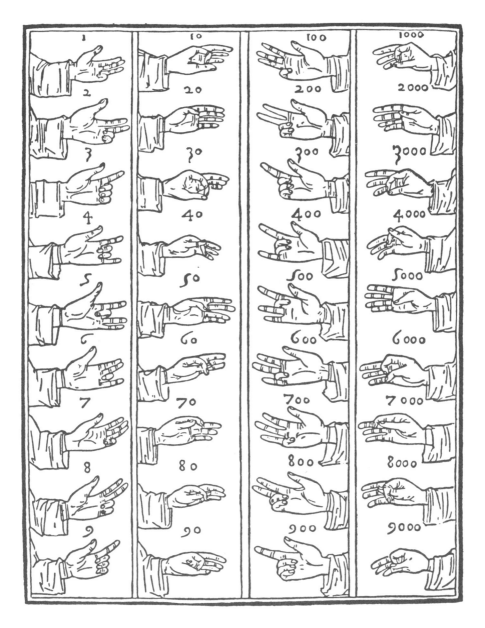

37

THE QUADRIVIUM
the qualities of quanta

Another word for a whole number is a *quantum* and the Quadrivium is an education in the behaviour of simple quanta. The purest study of quanta deals in factors, ratios, triangular, square, and cubic numbers, prime and perfect numbers, and the way numbers appear in sequences like the Fibonacci and Lucas sequences. We will meet many of these ideas as we go along, but dividing the unity of space and time also throws light on to the nature of quanta in these media.

For instance, opposite we see some of the limits placed by space on number. Allowing only perfect polygons there are three regular grids (*opposite top left*), five regular solids (*top right*), eight semi-regular grids (*center left*) and thirteen semi-regular solids (*center right*). These numbers, 3, 5, 8 and 13 are an interesting bunch, and we will meet them again in this book.

The numbers of music unfold as simple ratios between periods or frequencies (*lower, opposite*): 1:1 (unison), 2:1 (the octave), 3:2 (the fifth), and 4:3 (the fourth). The frequency of the fifth differs from that of the fourth as 9:8 (the value of the tone which gives rise to the scale).

The way number unfolds in space and time requires that we study the manifest cosmos, and the traditional subject of study here is the solar system. However, we could also add the beautiful simplicity of the periodic table, the quantum behavior of the subatomic realms, or the organisation of other natural phenomena with discrete elements.

Numerical facts of space and time are universal. They may or may not play the same tunes in the nearest intelligent galaxy, but they will agree that fifths sound lovely, and recognize five simple solids.

GNOMONS
ways of growing

Aristotle observed that some things suffer no change other than magnitude when they grow. He was describing the principle the Greeks referred to as 'gnomonic growth'. Originally a term for a carpenter's tool, a gnomon is defined as any figure, which, when added to another figure, leaves the resultant figure similar to the original. Contemplation of the gnomon leads to an understanding of one of nature's most common principles, growth by accretion. Structures such as bones, teeth, horns, and shells all develop in this way.

The ancients had a general fascination with patterns and progressions created by whole-number ratios. Examples are triangular, rectangular, square, and cubic numbers (*opposite top, and see too pages 358 and 367*); also Plato's *lambda*, or *lambdoma*, which produces the full range of musical ratios; and the proportional rectangles used in Greek design where each subsequent rectangle is built on the diagonal of the previous one (*opposite center*). The Fibonacci sequence is a more recent discovery, but relies on the same principle of gnomonic growth. The drawing below shows a cutaway cross section of the Aztec temple of Tenayuca, revealing five gnomic reconstructions, made every 52 years, when their calendar, inherited from the Mayans, was reset and all buildings renewed.

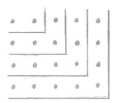

TRIANGULAR *numbers*
Here the sequence 1, 3, 6, 10
increases in a triangular fashion.

RECTANGULAR *numbers*
Here the sequence 2, 6, 12, 20
increases in a musical fashion.

SQUARE *and* CUBIC *numbers*
Here the square faces 1, 4, 9, 16
and the cubes 1, 8, 27, 64.

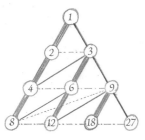

The LAMBDOMA
The heavier line shows the octave (2:1), while down
the other side numbers triple. Also present are the
fifth (3:2), fourth (4:3), and wholetone (9:8).

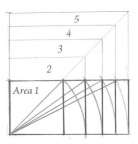

PROPORTIONAL RECTANGLES
Starting with a square side 1, each successive
rectangle is built on the diagonal of the previous
one to create squares of area 2, 3, 4, and 5.

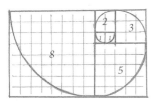

The GOLDEN SPIRAL.
Starting with a square we build new squares to create
a spiral of squares, which grows and grows by the magic
Fibonacci sequence 1, 1, 2, 3, 5, 8, 13, 21, 34, 55.

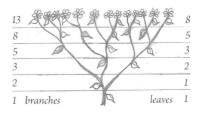

THE NUMBERS OF GROWTH
The Fibonacci sequence appears in many living
things. Here we see it in the number of leaves
and branches of a simple meadow daisy.

TIME AND SPACE
cosmology and manifest number

Looking around us, there are numbers that particularly manifest in the heavens around Earth. These are covered in detail in BOOK VI, but we will introduce a few here. There are, for instance, twelve full moons in a solar year, but the twelfth falls eleven days short of the end, which means that a twelve-moon year, like the Islamic calendar, slides slowly against the solar year, coming round again after 33 years, three elevens.

Other Sun-Moon marriage numbers are 18 and 19; as eclipses repeat after 18 years, and full moon dates repeat after 19 years. Stonehenge displays this as 19 stones in its inner horseshoe. Two full moons occur every 59 days, and Stonehenge records this in its outer circle of 30 stones, one of which is half-width, suggesting 29.5 days per moon.

Venus draws a fivefold pattern around Earth every eight years allowing us to draw an amazing diagram (*opposite center*). In those eight years there are almost exactly 99 full moons, nine elevens, the number of names or reflections of Allah in Islam. Jupiter draws a beautiful elevenfold pattern around Earth (*opposite top*).

The numbers of many longer cycles, such as the Great Year, or precession of the equinoxes, are also rich in secret qualities. Each great month, such as the Age of Pisces, or Aquarius, lasts 2,160 years, also the diameter of the Moon in miles. Twelve great months give the ancient Western value of 25,920 years for the whole cycle.

The ancient Maya were superb stargazers. Their calendar synchronized not just the Sun and Moon, but also Venus and Mars. They worked out that 81 (or $3 \times 3 \times 3 \times 3$) full moons occur exactly every 2,392 (or $8 \times 13 \times 23$) days, an astonishingly accurate gearing.

BABYLON, SUMER, AND EGYPT
early number systems

Around 3,000 BC the Sumerians developed the earliest writing we know of, and with it a base-60 number system (*see page 356*). A particularly useful number, 60 is divisible by 1, 2, 3, 4, 5, and 6.

Working in base-60 gives number patterns different from our modern base-10 system; a Sumerian clay tablet impressed with a cuneiform, 'wedge-shaped', stylus shows the 36 times table opposite. Their base-60 system survives today as our measurement of cycles and circles with 60 seconds in a minute, 60 minutes in an hour, or degree, and $6 \times 60 = 360$ degrees in a circle.

Ancient Egyptian numerals were made of characters standing for 1, 10, 100, and so on. An example of Egyptian arithmetic is their method of multiplication, which uses repeated doubling followed by selective addition to find the answer.

The ancient vision of number is a musical one in which every number inverts in the mirror of Unity, two becoming a half, three becoming a third, and so on. In base-60 this reciprocation is especially beautiful, as all multiples of 2, 3, 4, 5, and 6 become simple fractions. For example, 15 becomes a quarter. The Babylonians inherited and used this system to invoke their gods.

Egyptian fractions used a mouth hieroglyph (*below*), while fractions of volume were represented using the Eye of Horus.

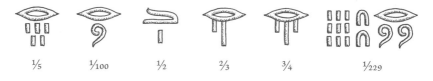

$1/5$	$1/100$	$1/2$	$2/3$	$3/4$	$1/229$

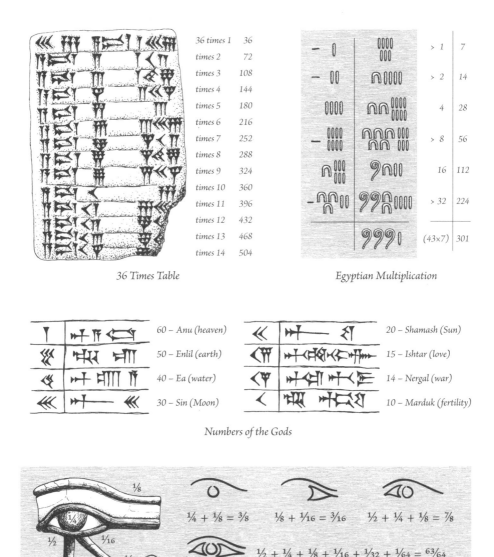

36 times 1	36
times 2	72
times 3	108
times 4	144
times 5	180
times 6	216
times 7	252
times 8	288
times 9	324
times 10	360
times 11	396
times 12	432
times 13	468
times 14	504

36 Times Table

	> 1	7
	> 2	14
	4	28
	> 8	56
	16	112
	> 32	224
	(43×7)	301

Egyptian Multiplication

		60 – Anu (heaven)
		50 – Enlil (earth)
		40 – Ea (water)
		30 – Sin (Moon)

		20 – Shamash (Sun)
		15 – Ishtar (love)
		14 – Nergal (war)
		10 – Marduk (fertility)

Numbers of the Gods

$\frac{1}{4} + \frac{1}{8} = \frac{3}{8}$ $\frac{1}{8} + \frac{1}{16} = \frac{3}{16}$ $\frac{1}{2} + \frac{1}{4} + \frac{1}{8} = \frac{7}{8}$

$\frac{1}{2} + \frac{1}{4} + \frac{1}{8} + \frac{1}{16} + \frac{1}{32} + \frac{1}{64} = \frac{63}{64}$

The Eye of Horus – Fractions of Volume

Ancient Asia
manipulating in tens

In China a written decimal system with 13 basic characters has been used for more than 3,000 years (*see page 356*). Another particularly beautiful system of writing numbers is the *suan zí* or *sangi* rod notation, complete with a small zero, used in China, Japan, and Korea in some form since at least 200 BC (*below*). Later, the famous Chinese abacus replaced rod-numeral counting boards. The speed of its operators, particularly in the Far East, is legendary, and it is still in widespread use today.

India has an ancient numerical tradition. Number is prominent in many of its scriptures, and Indian cosmology uses huge numbers rivaled today only by those of modern physics. Indian numerals originate with the Brahmi sytem of numerals, with 45 characters for the numbers 1 to 90,000. In time the speculations of Indian mathematicians required a new system combining the first nine number names with powers of ten. Rapid and elegant calculation techniques and the description of very large numbers resulted in some astonishing calculations.

The zero also emerged, to denote an empty decimal power without confusion. Indeed, it is from India that we received, via the Arabs, our modern decimal place value system.

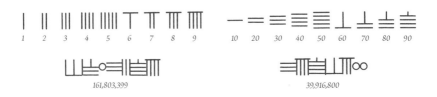

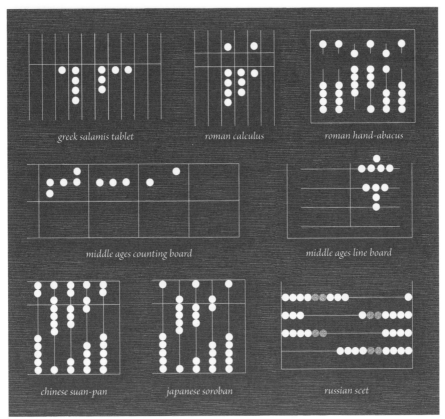

greek salamis tablet

roman calculus

roman hand-abacus

middle ages counting board

middle ages line board

chinese suan-pan

japanese soroban

russian scet

The number 9,360 on various counting boards

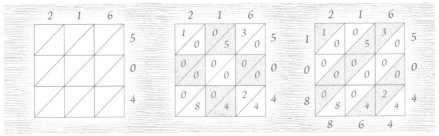

An Arabic use of Indian numerals – 216 multiplied by 504 equals 108,864

GEMATRIA

talking numbers and secret codes

The Phoenicians used a very neat twenty-two letter consonant alphabet to encode the sounds of their tongue. In time this script was adopted by most Mediterranean peoples and through its Latin variation came to be the alphabet that we use today.

Gematria uses letters as number symbols, so language becomes mathematics. Important canonical, geometrical, musical, metrological, and cosmological numbers are defined by many key terms in ancient texts. First appearing widely in ancient Greece, gematria was subsequently adapted to Hebrew and also to Arabic, where it is known as *abjad*. A simplified system also exists in all three languages using the same values without the zeros.

The example below shows two related phrases connected through an identical sum. It gives some idea of the magical and simultaneous resonance between words and numbers that any literate and numerate reader would have experienced, since for more than 1,000 years gematria was not merely an occult specialty but the standard way of representing numbers.

This secret science is still used today by mystics and sorcerers who use its connections between words, phrases, and number for their mystical significance and talismanic power.

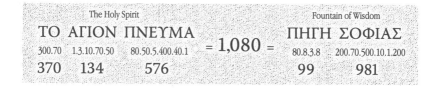

48

Archaic Phoenician		Greek			Hebrew		Arabic East | West		Value
ʾaleph	𐤀	alpha	A	α	aleph	א	ʾalif	ا	1
bet	𐤁	beta	B	β	bet	ב	ba	ب	2
gimmel	𐤂	gamma	Γ	γ	gimmel	ג	jim	ج	3
dalet	𐤃	delta	Δ	δ	dalet	ד	dal	د	4
he	𐤄	epsilon	E	ε	he	ה	ha	ه	5
waw	𐤅	digamma	F	ϛ	vov	ו	waw	و	6
zayin	𐤆	zeta	Z	ζ	zayin	ז	za	ز	7
ḥet	𐤇	eta	H	η	het	ח	ḥa	ح	8
ṭet	𐤈	theta	Θ	θ	tet	ט	ṭa	ط	9
yod	𐤉	iota	I	ι	yod	י	ya	ي	10
kaf	𐤊	kappa	K	κ	kof	כ	kaf	ك	20
lamed	𐤋	lambda	Λ	λ	lamed	ל	lam	ل	30
mem	𐤌	mu	M	μ	mem	מ	mim	م	40
nun	𐤍	nu	N	ν	nun	נ	nun	ن	50
samekh	𐤎	ksi	Ξ	ξ	samekh	ס	sin/ṣad	س ص	60
ʿayin	𐤏	omicron	O	o	ayin	ע	ʿayn	ع	70
pe	𐤐	pi	Π	π	pé	פ	fa	ف	80
ṣade	𐤑	qoppa	Ϙ	ϙ	tsade	צ	ṣad/ḍad	ض ص	90
qof	𐤒	rho	P	ρ	quf	ק	qaf	ق	100
resh	𐤓	sigma	Σ	σ	resh	ר	ra	ر	200
shin	𐤔	tau	T	τ	shin	ש	shin/sin	س ش	300
taw	𐤕	upsilon	Υ	υ	tav	ת	ta	ت	400
		phi	Φ	φ	kof	ך	tha	ث	500
		chi	X	χ	mem	ם	kha	خ	600
		psi	Ψ	ψ	nun	ן	dhal	ذ	700
		omega	Ω	ω	pé	ף	ḍad/ḍha	ظ ض	800
		san	ϻ	ϻ	tsade	ץ	dha/ghayn	ظ غ	900
							ghayn/shin	ش غ	1,000

The Greek system includes the disused letters *digamma* and *qoppa* in their original order and reinserts the disused *san* at the end. Likewise the Hebrew system uses five special 'end of word' letter forms to reach 900. In the Arabic system the letters for 60, 90, 300, 800, 900 & 1,000 differ in the West and the East of the Islamic world.

Magic Squares

when it all adds up

Magic squares are a fascinating way of arranging numbers, and there are whole books about them and their secret uses. The magic sum of any square is the same whichever line is added.

Seven magic squares are traditionally associated with the planets (*opposite*). The three-by-three square is Saturn's, and the squares increase by one order as they descend through each planetary sphere to reach the lunar nine-by-nine square. Elegant patterns of odd and even numbers occur in these squares (*shaded in the diagrams*). Each planet also has a magic seal based on the structure of its square, a useful code for wizards.

A magic square is an example of a permutation, ordering things in a set in a particular way. There are eight ways to sum to fifteen using three numbers from one to nine, and all eight ways are present in the three-by-three magic square.

Other totals found in magic squares are worth a second look. The Maya would surely have delighted in the fact that the eight-by-eight square has the magic sum of 13 × 20, while the solar line total of eleventy-one gives an ominous 666 square sum.

With gematria as an additional magical key, words and magic squares naturally interweave in the secret world of spells and other talismanic arts (*see example below*).

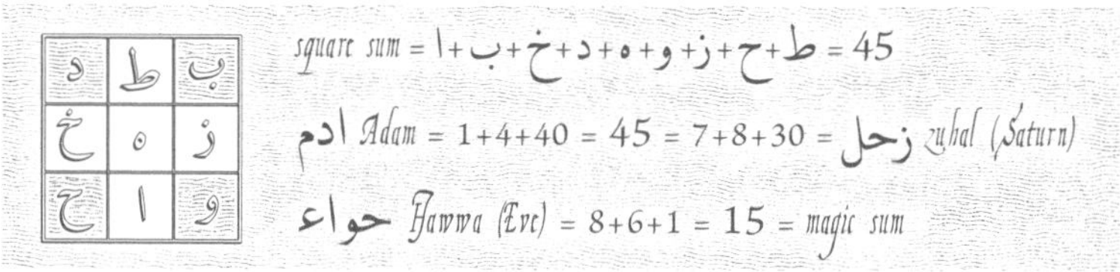

square sum = ١+ب+خ+د+ه+و+ز+ح+ط = 45

ﺍﺩﻡ *Adam* = 1+4+40 = 45 = 7+8+30 = زحل *zuhal (Saturn)*

حواء *Hawwa (Eve)* = 8+6+1 = 15 = *magic sum*

♄

4	9	2
3	5	7
8	1	6

magic sum 15
square sum 45

☽

37	78	29	70	21	62	13	54	5
6	38	79	30	71	22	63	14	46
47	7	39	80	31	72	23	55	15
16	48	8	40	81	32	64	24	56
57	17	49	9	41	73	33	65	25
26	58	18	50	1	42	74	34	66
67	27	59	10	51	2	43	75	35
36	68	19	60	11	52	3	44	76
77	28	69	20	61	12	53	4	45

magic sum 369
square sum 3,321

♃

4	14	15	1
9	7	6	12
5	11	10	8
16	2	3	13

magic sum 34
square sum 136

♂

11	24	7	20	3
4	12	25	8	16
17	5	13	21	9
10	18	1	14	22
23	6	19	2	15

magic sum 65
square sum 325

☿

magic sum 260
square sum 2,080

8	58	59	5	4	62	63	1
49	15	14	52	53	11	10	56
41	23	22	44	45	19	18	48
32	34	35	29	28	38	39	25
40	26	27	37	36	30	31	33
17	47	46	20	21	43	42	24
9	55	54	12	13	51	50	16
64	2	3	61	60	6	7	57

☉

6	32	3	34	35	1
7	11	27	28	8	30
19	14	16	15	23	24
18	20	22	21	17	13
25	29	10	9	26	12
36	5	33	4	2	31

magic sum 111
square sum 666

♀

magic sum 175
square sum 1,225

22	47	16	41	10	35	4
5	23	48	17	42	11	29
30	6	24	49	18	36	12
13	31	7	25	43	19	37
38	14	32	1	26	44	20
21	39	8	33	2	27	45
46	15	40	9	34	3	28

MYTH, GAME, AND RHYME

numbers we grow up with

Some of our earliest experiences with number occur by way of games, rhymes, stories, and cultural myths, many of which are treasure troves of hidden numerical relationships.

Ancient forms of language were regularly number-based, so in poetry we find triplets (three lines of verse), quatrains (verses of four lines), pentameters (lines with five stressed syllables), hexameters (lines with six stressed syllables), and haiku (a three line poem of seventeen syllables: five, seven, then five; *compare with the 17-note scale page 196*).

Games, like myths and stories, can store information. The sum of a pack of playing cards, counting jack, queen and king as 11, 12, and 13, is 364, which with the joker produces 365, the number of days in a year. The eighteens and nineteens of the Chinese game of Go echo the cycles of the Sun and the Moon (*see page 42*). These ancient games reflect eternal principles, suggesting larger cosmic games, also with number at their center.

Many games are dependent upon number for their structure and rules. Imagine a game of poker played by people who couldn't count higher than three! Below are two examples of knight's tours from chess, both of which produce magic squares when numbered in sequence.

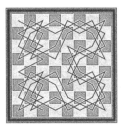 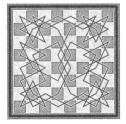

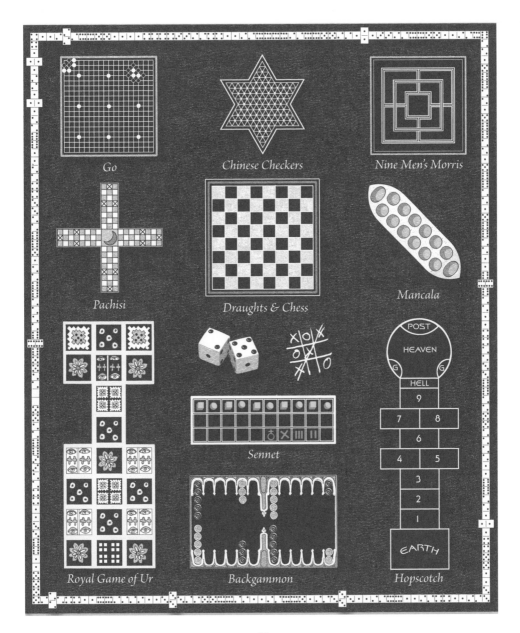

Go

Chinese Checkers

Nine Men's Morris

Pachisi

Draughts & Chess

Mancala

Royal Game of Ur

Sennet

Backgammon

Hopscotch

MODERN NUMBERS
the dawn of complexity

When the ancient Greeks proved that the diagonals of squares could not be expressed as fractions, it is said to have caused a crisis in their ranks, a little like the terror still experienced today by many people faced with a square root symbol, √.

The last 400 years of human thought have transformed our conception of number. After the adoption of Indian numerals and the zero, the next piece of witchcraft was the invention of *negative* numbers, creating a number line which vanished in two directions. Negative numbers were helpful but created a conundrum: square a negative number and it becomes positive—so what are the square roots of negative numbers? Mathematicians realized that there was another entire number line, of the square roots of negative numbers, which they called *imaginary* numbers, labeled today with an *i* (so *i* is the square root of minus one). Numbers today live on a number plane, with a real part and a complex part. Interestingly, it is the play between imaginary and real numbers that effortlessly produces the beautiful complexity of fractals and chaos theory, models of the recursive shapes and processes we find all around us in nature.

With the decimal system we use today, we can describe numbers like π, or *pi*, the ratio between a circle's circumference and its diameter, with great accuracy. However, some of the most beautiful objects in modern mathematics simply employ repeated fractions which would have been familiar to the ancients. These capture the complex essence of square roots, the *Golden Section* φ or Φ, *pi* π, and the exponential growth function, *e*.

$$\phi = \frac{\sqrt{5}+1}{2}$$

$$\sqrt{2} + \sqrt{3} + \sqrt{5} + \phi \approx 7$$

$$\pi \approx 6/5\,\phi^2$$

$$\sqrt{2} = 1.41421356237\ldots$$
$$\sqrt{3} = 1.732050807569\ldots$$
$$\phi = 1.61803398875\ldots$$
$$\sqrt{5} = 2.2360679775\ldots$$
$$e = 2.71828182846\ldots$$
$$\pi = 3.14159265359\ldots$$

$$\sqrt{2} = 1 + \cfrac{1}{2 + \cfrac{1}{2 + \cfrac{1}{2 + \cfrac{1}{2 + \cdots}}}}$$

$$\sqrt{3} = 1 + \cfrac{1}{2 + \cfrac{1}{1 + \cfrac{1}{2 + \cfrac{1}{2 + \cfrac{1}{2+\cdots}}}}}$$

$$\sqrt{5} = 2 + \cfrac{1}{4 + \cfrac{1}{4 + \cfrac{1}{4 + \cfrac{1}{4 + \cdots}}}}$$

$$\phi = 1 + \cfrac{1}{1 + \cfrac{1}{1 + \cfrac{1}{1 + \cfrac{1}{1 + \cdots}}}}$$

$$\sqrt{-1} = i$$

$$e^{i\pi} + 1 = 0$$

$$V - E + F = 2$$

$$\frac{\pi}{4} = \frac{1}{1} - \frac{1}{3} + \frac{1}{5} - \frac{1}{7} + \frac{1}{9} - \frac{1}{11} + \frac{1}{13} \cdots$$

$$e^x = 1 + x + \frac{x^2}{2!} + \frac{x^3}{3!} + \frac{x^4}{4!} + \frac{x^5}{5!} + \cdots$$

$$e = 1 + 1 + \frac{1}{2!} + \frac{1}{3!} + \frac{1}{4!} + \frac{1}{5!} + \cdots$$

$$\sin x = x - \frac{x^3}{3!} + \frac{x^5}{5!} - \frac{x^7}{7!} + \cdots$$

$$\cos x = 1 - \frac{x^2}{2!} + \frac{x^4}{4!} - \frac{x^6}{6!} + \cdots$$

$$r = \sqrt{x^2 + y^2}$$
$$y = r\sin\theta$$
$$x = r\cos\theta$$
$$y = x\tan\theta$$

ZERO

nothing left to say

Zero has been left until last, because in a sense it is not actually a number at all, just a mark representing the absence of number. It is perhaps for this reason, and the horror many theologians had of it, that nothing took such a long time to emerge as something at all, and in quite a few sensible cultures it never did.

A symbol for zero has been invented independently at least three times. The Babylonians in 400 BC started using the shape of two wedges pressed into clay to act as an 'empty place' marker in their sexagesimal numerals, 'no number in this column'. On the other side of the world, and nearly a thousand years later, the Mayans adopted a seashell symbol for the same function.

The circular form that 'nothing' assumed under the Indians reflected the indentation left in sand when a pebble used for counting is removed. Thus our modern zero, inherited from the Indians, began as the visible trace of something no longer there.

Like one, zero probes the borderline between absence and presence. In early Indian mathematical treatises it is referred to as *Sunya*, meaning 'void', calling to mind the abyss, the ultimate unknowable, the pregnant ground of all being.

It is perhaps appropriate that our zero takes the form of a circle, itself a symbol of one, and that our one takes the form of a short line between two points. As acknowledged in gematria, each number already contains the seed of its successor within it, and the symbols for zero and one strangely combine to create the Golden Symbol φ, a fitting thought with which to end this book.

BOOK II

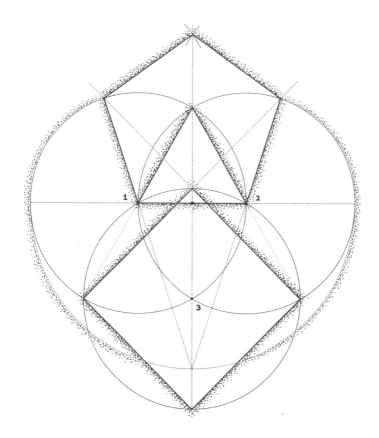

SACRED
GEOMETRY

Miranda Lundy

Seats in Norwoods chantry, in Milton church, Kent.

INTRODUCTION

Sacred Geometry charts the unfolding of number in space. It differs from mundane geometry purely in the sense that its moves, concepts and products are regarded as having symbolic value and meaning. Thus, like good music, the study and practice of geometry can facilitate the evolution of the soul. As we shall see, the basic journey is from the single point, into the line, out to the plane, through to the third dimension and beyond, eventually returning to the point again, watching what happens on the way.

These pages, BOOK II of *Quadrivium*, cover the elements of two-dimensional geometry—the unfolding of number on a flat surface. The three-dimensional geometrical story is then told in BOOK III. This material has been used for a very long time indeed as one introduction to metaphysics. Like the elements of its sister subject, music, it is an aspect of revelation, a bright indisputable shadow of Reality and a creation myth in itself.

Number, Music, Geometry, and the study of patterns in the Heavens are the four great Liberal Arts of the ancient world dealing with quanta, or whole numbers. These simple universal languages are as relevant today as they have always been, and may be found in all known sciences and cultures without disagreement. Indeed, one would expect any reasonably intelligent three-dimensional being anywhere in the universe to know about them in much the same way as they are presented here.

Just above the entrance to Plato's Academy was a sign: "Let none ignorant of geometry enter here." So let's do some research.

POINT, LINE, AND PLANE
none, one and two dimensions

Begin with a sheet of paper. The point is the first thing that can be done. It is without dimension and is not in space. Without an inside or an outside, the point is the source for all which now follows. The point is represented (*below*) as a small circular dot.

The first dimension, the line, comes into being as the One emerges into two principles, active and passive (*below right*). The point chooses somewhere outside of itself, a direction. Separation has occurred and the line comes into being. A line has no thickness, and it is sometimes said that a line has no end.

Three 'ways' now become apparent (*opposite*):

i) With one end of the line stationary, or passive, the other is free to rotate and describe a circle, representing Heaven.

ii) The active point can move to a third position equidistant from the other two, thus describing an equilateral triangle.

iii) The line can produce another which moves away until distances are equal to form a square, representing Earth.

Three forms, circle, triangle and square have manifested. All are rich in meaning. Our journey has begun.

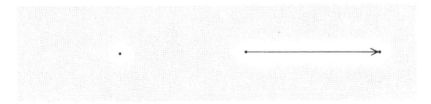

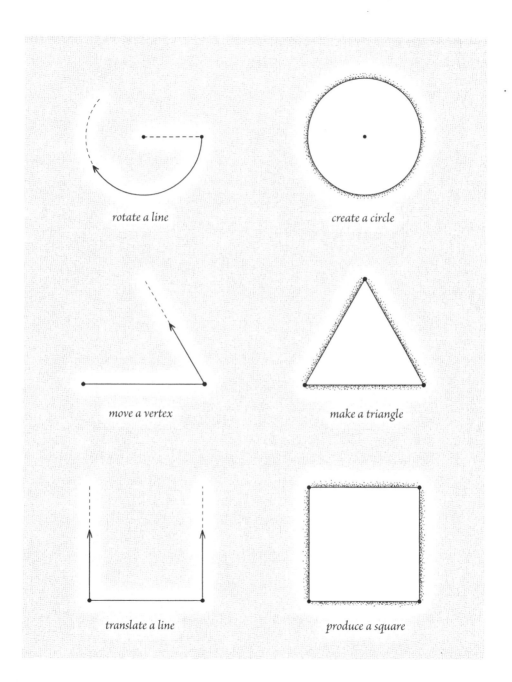

rotate a line

create a circle

move a vertex

make a triangle

translate a line

produce a square

SPHERE, TETRAHEDRON, AND CUBE
from two to three dimensions

Although this book concerns itself primarily with the plane, the three 'ways' of rotation, movement of a *vertex*, and translation of an object, are here taken one step further, into three dimensions (*opposite*):

i) The circle spins to become a sphere. Something circular remains essentially circular (*top row*).

ii) The triangle produces a fourth point at an equal distance from the other three to produce a tetrahedron. One equilateral triangle has made three more (*central row*).

iii) The square lifts a second square away from itself until another four squares are formed and a cube is created (*lower row*).

Notice how the essential division into circularity, triangularity and squareness from the previous page is preserved.

The sphere is a symbol of the cosmos and the totality of manifest creation. Very large and very small things in nature tend to be spherical. Einstein discovered that a point in four dimensions (i.e. you here and now) is a sphere expanding at the speed of light, and all we can see of the entire universe is inside an event-horizon sphere. The cube represents the Earth.

The sphere possesses the smallest surface area for its volume of any possible three-dimensional solid whereas, amongst regular solids, the tetrahedron is the opposite.

A tetrahedron is in fact hiding in a cube—if you draw a single diagonal line on every face of a cube so that they join at the corners, you will have defined the edges of a tetrahedron. Try it!

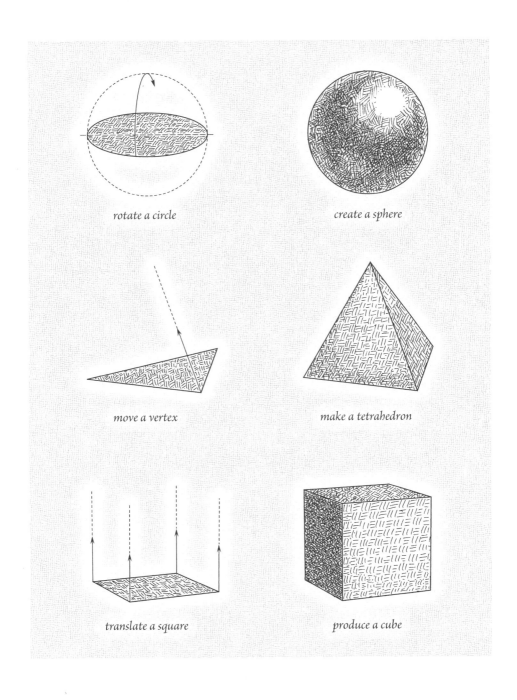

rotate a circle

create a sphere

move a vertex

make a tetrahedron

translate a square

produce a cube

ONE, TWO, AND THREE

playing with circles

Get a ruler, compass, something to draw with and something to draw on. Draw a horizontal line across the page. Open the compass and place the point on the line. Draw a circle (*top*).

Where the circle has cut the line, place the compass point and draw another circle, leaving the compass at the same opening as before. When one circle is drawn over another like this so that they pass through each others' centers, then an important almond shape, the *vesica piscis*, literally 'fish's bladder', is formed. It is one of the first things that circles can do. Christ is often depicted inside a *vesica*. Two equilateral triangles have been defined (*opposite center*).

A third circle can be added to the line as before, normally on the other side of the forming circle, this simple act defining all six points of a perfect hexagon (*lower, opposite*). Alternatively, the third circle can be added as shown below to produce an elegant triangular form.

Circles thus effortlessly produce perfect triangles and hexagons.

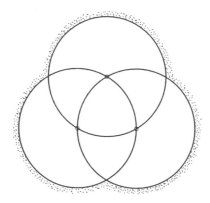

SIX AROUND ONE

or twelve or even eighteen

The six points of the hexagon give rise to the flower-like pattern shown below. Alternatively it can be drawn by 'walking' a circle around itself—something most children have done at school, whether under instruction or just playing with a compass.

We are now seeking the lower diagram opposite, and need the centers of the six outer circles. One way is to extend the flower, lightly drawing the six circles shown dashed in the top diagram above, to give us the six centers. Otherwise we can draw straight lines as shown in the lower diagram. Both ways work.

We can now see that six circles fit around one. We can push glasses, coins or tennis balls together to see it, yet it is extraordinary really. 'Six around one' is a theme which the Old Testament of the Bible opens on, with the six days of work and the seventh day of rest. There is indeed something very sixy about circles.

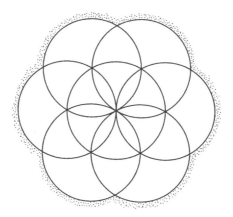

TWELVE AROUND ONE

how to draw a dodecagon

As one produces six, so six produces twelve. Here the arms of a six-pointed star extend from the flower to intersect the outer rims of the six circles. Beautifully, this forms a perfect overall division of space into twelve parts (*shown opposite*). The twelve-sided polygon is called a dodecagon, which means literally 'twelve sided'.

The dodecagon is also made from six squares and six equilateral triangles fitted around a hexagon—can you see them all opposite? In addition, the shape divides into its factors, three, four and six, as four triangles, three squares, and two hexagons (*lower, opposite row*).

Shown below is the three-dimensional version of the same story. A ball naturally fits twelve others around it so that they all touch the center and four neighbours. You see this arrangement in apples and oranges in every market stall. The shape made is called the cuboctahedron and is closely related to the tetrahedron and cube we saw on page 67. Most crystals grow along these lines.

Twelve is the number which fits around one in three dimensions in the same way that six fits around one in two dimensions. The New Testament is a story of a teacher surrounded by twelve disciples.

THE FIVE ELEMENTS
a brief foray into the third dimension

Although covered in depth in BOOK III of this volume, it is worth mentioning here that there are just five regular three-dimensional solids. Each has equal edges, every face is the same perfect polygon and every point is the same distance from the center. Known as the five Platonic Solids, they were recognized in the British Isles two thousand years before Plato—4,000 year-old carved stone sets of them have been found at stone circles in Aberdeenshire, Scotland (*below, from Critchlow*).

The first solid is the tetrahedron, with four vertices and four faces of equilateral triangles, traditionally representing the element of Fire. The second solid is the octahedron, made from six points and eight equilateral triangles, representing Air. The Cube is the third solid, eight vertices and six square faces, representing Earth. The fourth is the icosahedron, with twelve points and twenty faces of equilateral triangles, the element of Water. The last, and fifth, element is the dodecahedron, which has twenty vertices, representing the mysterious fifth element of Aether.

Notice how beautiful the dodecahedron is, and how it is made of twelve pentagons, perfect five-sided shapes.

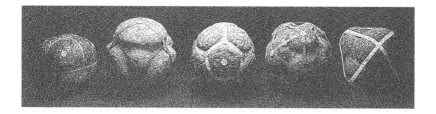

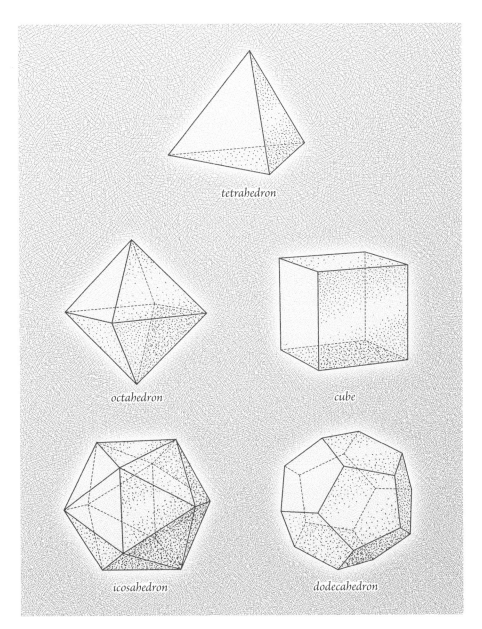

tetrahedron

octahedron

cube

icosahedron

dodecahedron

CIRCLING THE SQUARE

the marriage of heaven and earth

The circle is the shape traditionally assigned to the Heavens, and the square to the Earth. When these two shapes are unified by being made equal in area or perimeter we speak of 'squaring the circle', meaning that Heaven and Earth, or Spirit and Matter, are symbolically combined, or married. Fivefold Man exists between sixfold Heaven and fourfold Earth and Leonardo da Vinci's image (*opposite*) also shows how a man's span equals his height, that this measure equals seven of his feet and other important ratios.

As we saw earlier (*page 33*), the Earth and the Moon square the circle, for if the Moon (diameter 3) is drawn down to the Earth (diameter 11) then a heavenly circle through the Moon (*dotted, below center*) has radius 7, and so circumference 44, the same as the perimeter of the square around the Earth. This works because π, which relates the circumference of a circle to its diameter is practically ²²⁄₇. In Leonardo's image the Moon would fit above the man's head.

Also shown (*below left and right*) is a simple construction for a square using ruler and compass. Octagons soon follow.

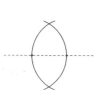

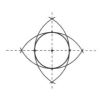

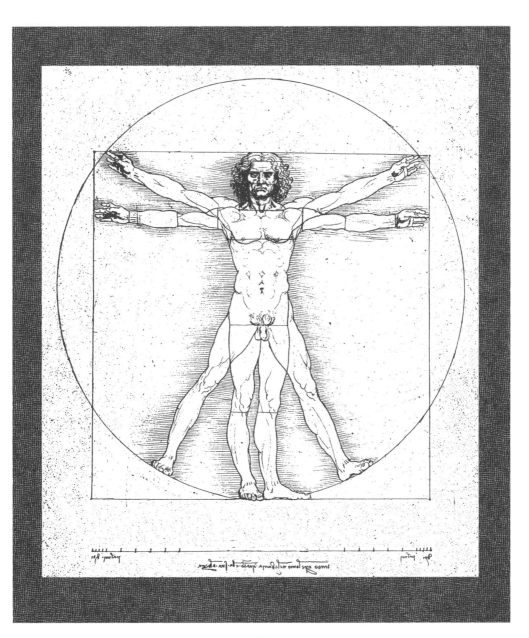

THE CANON
3, 7 and 11

The squaring of the circle by the Moon (size 3) and the Earth (size 11) is also manifest in the geometry of a double rainbow, whose two beautiful bows at 41.5° and 52.5° precisely draw the same diagram (*below, from Martineau*)—a marriage of Heaven and Earth indeed!

A portal door, of Gerum Church in Gothland, Sweden (*opposite*) encodes 3 by 11. Three elevens is thirty-three and Irish and Norse myths abound with tales of 33 warriors. Jesus dies and is resurrected aged 33, and the Sun takes 33 years for a perfect repeat sunrise. Seven also works with both 3 and 11. It is an old secret that the Earth's tilt, often hidden in sacred art as the tilt of a holy head (the Virgin Mary's or the Buddha's), is easily produced as the diagonal of a rectangle 3 wide and 7 high. Finally $^{11}/_7$ is the ancient Egyptian value for half of π.

The Sandreckoner's diagram (*opposite*) is a unique way of dividing a rectangle's edge into harmonic fractions (*after Malcolm Stewart*).

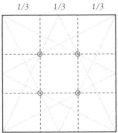

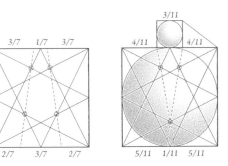

3/7 1/7 3/7

2/7 3/7 2/7

3/11

4/11 4/11

5/11 1/11 5/11

Above and below: The Sandreckoner's diagram. Simply by joining corners to the centers of sides, a square's edge may be exactly divided into 3, 4, 5, and above, 7 and 11 equal parts, which makes it an extremely useful device. The initial lines also produce a plethora of whole number lengths, areas, and shapes, including a multiude of Pythagorean 3-4-5 triangles at various scales.

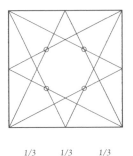

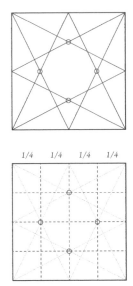

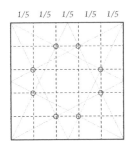

1/3 1/3 1/3

1/4 1/4 1/4 1/4

1/5 1/5 1/5 1/5 1/5

PYRAMID PI
a marriage of everything

There is perhaps no more famous a geometric object on Earth than the Great Pyramid at Giza in Egypt with its strange passages and enigmatic chambers. The five diagrams opposite show:

1. The square of the height is equal to the area of each face.
2. The *Golden Section* in the pyramid, Φ = 1.618 (*see page 86*).
3. *Pi* in the pyramid. Pi, or π, defines the ratio between a circle's circumference and its diameter (3.14159...).
4. The pyramid squaring the circle (*see page 76*).
5. A pentagram defining a 'net' for the pyramid - cut and fold!

Geometry means 'Earth-measure'. The Pyramid functions as a ridiculously accurate sundial, star observatory, land surveying tool and repository for weights and measures standards. Written into the design are highly accurate measurements of the Earth, detailed astronomical data and these simple geometric lessons.

A 3-4-5 triangle fits the shape of the King's Chamber (*below*) and also gives the angle of slope of the second pyramid at Giza. Halfway between the two slopes is 51.4°, one seventh of a circle.

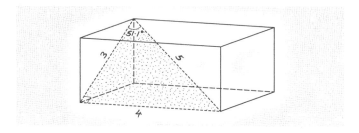

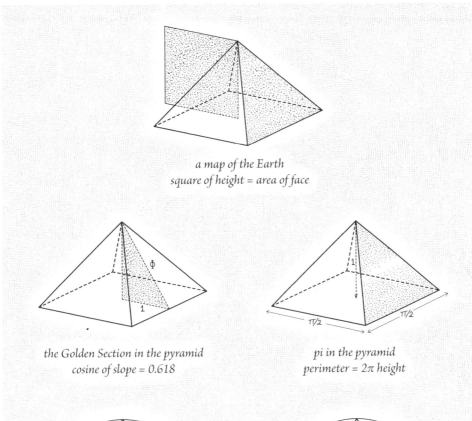

a map of the Earth
square of height = area of face

the Golden Section in the pyramid
cosine of slope = 0.618

pi in the pyramid
perimeter = 2π height

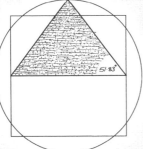

circumference of circle on height
= perimeter of pyramid base

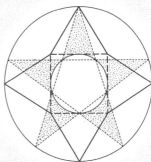

constructing the pyramid from a
pentagram drawn in a circle

HALFLINGS AND THIRDLINGS
defined by triangles and squares

An equilateral triangle (*opposite top left*), or two nested squares (*opposite top right*) both do the same thing—the circle inside each of these figures is exactly half the size of the surrounding circle. This is a geometrical image of the musical octave, where a string-length or frequency is halved or doubled.

Appropriately, the three-dimensional equivalent of the triangle, the tetrahedron, defines the next fractional proportion, one third, as the ratio of the radius of the innermost sphere to that of the containing sphere (*opposite, bottom left*). Two nested cubes, or two nested octahedra, or an octahedron nested in a cube (*opposite, bottom right*) all produce one third too. The geometric third is musically equal to an octave plus a fifth in harmonic notation. Thus two dimensions quickly define a *half*, and *three* dimensions a *third*.

A close but not perfect marriage is between five and eight (*below*), whose geometries often play with one another. In both these diagrams the inner circle could be the size or orbit of the planet Mercury if the outer circle is taken as being the size or orbit of the Earth.

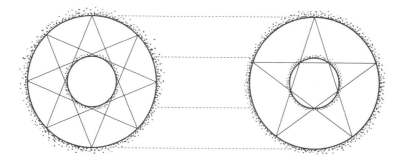

ad triangulum
an equilateral triangle defines two
circles, one half the size of the other

ad quadratum
two nested squares define two circles,
again one half the size of the other

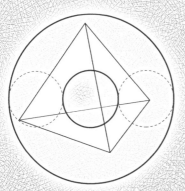

ad tetratum
the sphere inside a tetrahedron is one
third of the size of the sphere outside

ad cuboctum
a cube and octahedron again define
an insphere one third of the outsphere

THE SHAPES OF SOUNDS
and three quarters

Geometry is 'number in space', music is 'number in time'. As we shall see in BOOK IV, basic musical intervals are simply elementary ratios, 1:1 (unison), 2:1 (the octave), 3:2 (the fifth), 4:3 (the fourth) and so on. The difference between the fourth and the fifth, which works out at 9:8, is the value of one tone. Musical *intervals*, like geometrical proportions, always involve two elements in relationship, two string-lengths, two periods (lengths of time) or two frequencies (beats per length of time). Simple ratios sound and look beautiful.

We can see musical intervals as shapes by swinging a pen in a circle at one frequency, and a table in an opposite circle at another frequency, the device being called a *harmonograph*. Shown opposite are two patterns from near-perfect intervals. The octave (*upper*) draws as a triangular shape, the fifth (*lower*) a pentagonal form.

In the spirit of the previous page, two octaves, or a quarter, can be exactly defined by two triangles, four squares, or, intriguingly, by a pentagon in a pentagram (*below*).

THE GOLDEN SECTION
and other important roots

A pentagram inside a pentagon is shown opposite. A simple knot, carefully tied in a ribbon or strip of paper and pulled tight and flattened out makes a perfect pentagon. Try it some time!

In the main diagram opposite you can see that pairs of lines are each dashed in different ways. The length of each such pair of lines is in the *Golden Section* ratio, $1 : \phi$, where ϕ (pronounced *'phi'*) can be either 0.618 or 1.618 (more exactly 0.61803399…). The books in this volume use the lower case ϕ for 0.618 and the upper case Φ for 1.618.

Importantly, ϕ divides a line so that the ratio of the lesser part to the greater part is the same as the ratio of the greater part to the whole. No other proportion behaves so elegantly around unity. For instance, $1 \div 1.618$ is 0.618, and $1.618 \times 1.618 = 2.618$. So one divided by Φ equals ϕ (or Φ minus one), and Φ multiplied by Φ equals Φ plus one!

The Golden Section is one of three simple proportions found in the early polygons (*lower, opposite*). With edge-lengths 1, a square produces an internal diagonal of $\sqrt{2}$ (the square root of two), a pentagram Φ, and a hexagon $\sqrt{3}$ (the square root of three). Many familiar objects from cassettes to credit cards and Georgian front doors are Φ rectangles. $\sqrt{2}$ and $\sqrt{3}$ are found widely in crystals, while Φ appears predominantly in organic life, possibly due to the flexible icosahedral nature of water and other liquids. All three geometric proportions are employed in good design, along with harmonic ratios.

Neighbouring terms in the Fibonacci Series: 1, 1, 2, 3, 5, 8, 13, 21, 34, 55 … (adding each pair of numbers to get the next) approximate ϕ with increasing accuracy. For the keen, $\phi = \frac{1}{2}(\sqrt{5}-1)$ and $\Phi = \frac{1}{2}(\sqrt{5}+1)$.

SOME SPECIAL SPIRALS
and how to draw them

Spirals are marvellous forms which nature uses at every scale. Three have been selected for this book, all of which give the impression of a spiral from multiple arcs of circles.

The first is the Greek Ionic volute shown top left. This is quite hard to draw and the secret lies in the small 'key' shown above it. The dotted lines in the main drawing show the radii of the arcs and give clues to the centers. It's not as hard as it looks!

Regular spirals such as the one shown top right also need a key. This can simply be two dots (the easiest), a triangle, a square, a pentagon, or a hexagon (*as shown*). The more points you have the more perfect the spiral will be. Here's how to do one with just two dots. Draw two dots quite close together and draw a semicircle centered on one starting from the other. Now, keeping the pen in the same place, open the compass a bit wider, moving the point to the other dot and continue in the same direction, drawing another semicircle. Repeat this a few times and a spiral will appear. It sounds harder than it is—if you try it you will soon get the idea. The bigger the key the wider the coils. Now look at the Ionic volute key again—can you see what is happening?

The bottom picture shows a Golden Section spiral, one of the family of exponential spirals which are common throughout the natural world. A Golden Section rectangle has the special property that removing a square from it produces another Golden Section rectangle, and the Golden Section spiral is formed by removing successive squares and filling each with a quarter arc.

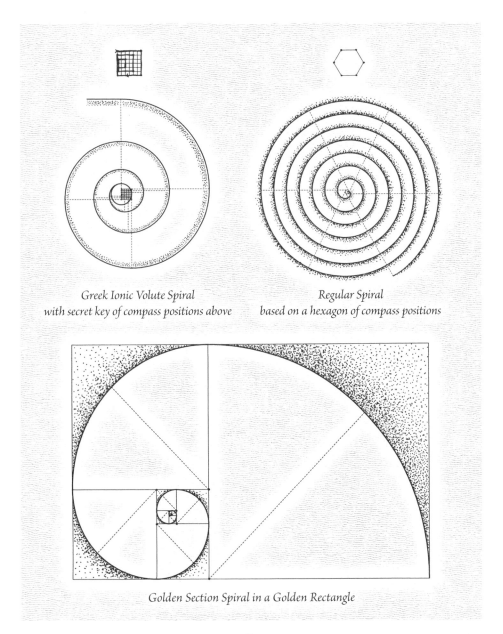

Greek Ionic Volute Spiral
with secret key of compass positions above

Regular Spiral
based on a hexagon of compass positions

Golden Section Spiral in a Golden Rectangle

HOW TO DRAW A PENTAGON

and a golden section rectangle

The method of construction of a pentagon shown opposite is perfect and is from the *Almagest* of Ptolemy (d. *ca.* 168 AD).

Draw a horizontal line with a circle on it. Keeping the compass opening fixed, place the point at ‹1› and draw the vesica through the center of the circle. Now open the compass wide and draw arcs from ‹1 › and ‹2› to cross above and below the circle. Use a straight edge to draw the vertical through the center of the circle. Next draw the vertical through the vesica to produce ‹3›. With the point of the compass at ‹3 › swing an arc down from ‹4› at the top of the circle to give ‹5›. With the point at ‹4› swing through ‹5› to give two points of the pentagon. With the point of the compass on these new points in turn, swing from the top to find the last two points of the pentagon.

A Golden Section rectangle, widely used in painting and architecture, is constructed from the mid-point of the side of a square (*below*).

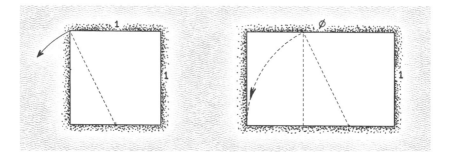

an ancient method for the construction of a
perfect PENTAGON - numbered points
represent compass positions

a decagon (10 sides) may be easily
constructed from a pentagon -
ten pentagons precisely fit
around a decagon

the pentagon and its starry twin the
pentagram are both symbols of water,
each molecule of which is the corner of a
pentagon, and of the life-force itself

the number five represents the sacred
marriage of two (female) and three
(male) and so symbolizes reproduction
and the magical healing arts

THE HEPTAGON
seven out of three

Divide a circle into six and draw the primary equilateral triangle. Find the midpoints ‹1› and ‹2› of the triangle's two upper arms and drop two lines down to give two points ‹3› and ‹4› on the base of the triangle, and two on the bottom of the circle. Finally, from the top, swing through the four points on the triangle to give the last four points of the seven on the circle.

Although it is impossible to draw a precise heptagon using ruler and compass alone, you can do it perfectly using seven equal rods or matchsticks (*shown below left*). This wedge is an *exact* fourteenth of a circle, so you need two of them for a one seventh division. More ancient rough solutions use a cord with either six knots or in a loop with thirteen (*below center and right*).

The ancient builders were amazing surveyors. Avebury stone circles in England are positioned *exactly* at latitude 51.4°, one seventh of a circle up from the equator. Luxor in Egypt is at a latitude exactly halfway between Avebury and the equator. Mecca meanwhile is at the northern Golden Section latitude between the two poles.

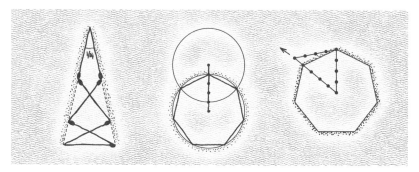

a secret method for the construction of a
near-perfect HEPTAGON starting with an
equilateral triangle inside a circle

introduces no compound errors as each
vertex is defined independently - it is not
possible to construct a perfect heptagon
using ruler and compass alone

seven symbolizes the virgin, having
little to do with the other low numbers,
and being complete in herself - many
traditions revere seven as especially holy

the seven notes of the scale, seven days
of the week, seven heavenly bodies of
antiquity and seven bodily chakras all
indicate the ancient sanctity of seven

THE ENNEAGON
nines and magic lozenges

The construction shown opposite divides a circle into a near-perfect nine from an initial six-pointed star using three centers.

The digits of many special numbers sum to nine: 2,160 or 7,920 for instance, the diameters of the Moon and the Earth in miles; or 360 and 666, and pentagonal angles like 36, 72, and 108. In fact, all multiples of nine add up to nine. Nine is three times three, or three squared. Many tribal cultures speak of nine worlds, or nine dimensions.

The golden Bush Barrow lozenge found near Stonehenge (*below left*) has internal angles of 80° and 100°, suggesting nine-fold geometry. Sunrises and sunsets at the latitude of Stonehenge vary over 80°, and moonrises and moonsets over 100°, so this was a useful object.

Not for beginners is the obscure fact that a sphere-point enneagon (or nine coins arranged in a perfect nine-sided figure) can contain two more spheres which exactly touch (*below right*).

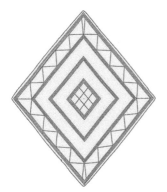

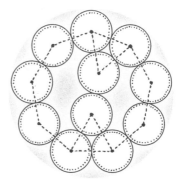

John Michell's method for the construction
of a near-perfect ENNEAGON starting with a
hexagram inside a circle

a simple and memorable
method accurate enough for
most practical purposes

the symbolism of nine is laden with
initiatory veils with stories of nine worlds
or realms in many shamanic traditions

nine is the square of three and
along with eight, the cube of two, is very
important in Eastern cosmologies

RABATMENT
and the rule of three

Painters have a whole bag of secret tricks to help them produce the perfect picture. Students of composition are taught 'the rule of three', where a canvas is divided in three horizontally and vertically, into nine small versions of the original rectangle (*below left*). The four intersections produced are excellent places to choose as focus points in the design and are used by many artists. By contrast, items placed on center-lines seem contrived in the composition, too obvious.

Another trick is to draw a square within your rectangle and use the lines produced as focal axes (*below center*). This is called rabatment. In Golden Section rectangles the space left over from a square is another Golden rectangle. The process may be continued indefinitely.

Occult centers (*below right*) are found by using the diagonals of a rectangle, and right-angled triangles with the other corners.

Dividing a line into 2 or 3 parts uses the first few steps of the Fibonacci sequence, 1, 2, 3, 5, 8, 13, 21, and so on, where adjacent terms home in on the Golden Section 0.618. Some painters use ⅗ divisions or even the Golden Section itself. Opposite we see two wonderful examples of rabatment at play. Botticelli uses a Golden Section rectangle reduced by rabatment in stages to compose his painting, while Grimshaw uses a rectangle between Φ and √2, with halves and Golden lines as guides.

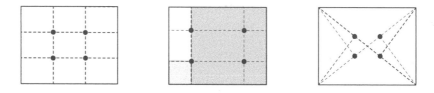

Above: Botticelli's Birth of Venus. A Golden Section rectangle with interior squares define the horizon, spine, navel and other features. Below: John Atkinson Grimshaw's Iris fairy is positioned using center lines and golden divisions. The fairy exists in half the painting.

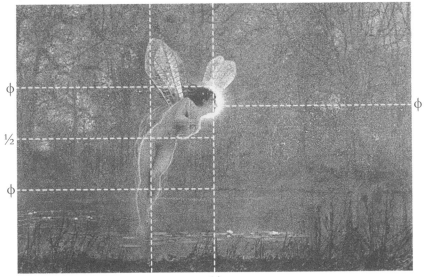

SIMPLE TILINGS

repeating patterns over an infinite surface

A regular tiling (or *tessellation*) of the plane occurs when the same regular polygon is used to fill the plane, leaving no spaces. Only three of these are possible (*shown below*). A *semi-regular* tiling allows for more than one type of polygon but insists that each vertex is the same. For example, in the central pattern opposite every vertex is a meeting of two hexagons and two triangles. Eight semi-regular tilings are possible and all are shown opposite (though the top left and top right grids opposite are left and right-handed versions of each other and count as one).

Some designs can be filled in further. As shown on page 73, dodecagons are just made of hexagons, triangles and squares, and hexagons are simply collections of triangles. As we shall soon see, triangles and squares can go on to do the most amazing things together. What of the other regular polygons? Octagons only tile with squares (*opposite top center*). Pentagons do not fit together happily on the plane, preferring the third dimension (*see pages 74–75*).

Heptagons and enneagons stand aloof.

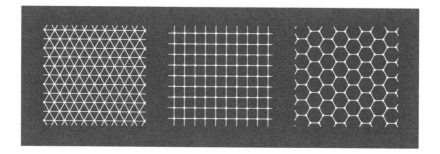

FURTHER TILINGS
further fun in the bathroom

There are twenty demi-regular tilings (where two vertex situations are permitted) and most of these, and a few other interesting ways to tile the plane, are shown on these two pages.

These tessellations form the basis for pattern construction in many traditions of sacred and decorative art across the world. They can be found underlying Celtic and Islamic patterns and in the natural world they appear as crystal and cellular structures. William Morris used them widely for his repeat wallpaper and fabric designs. Their uses are limited only by your imagination!

On the next page we see one of these grids put to use.

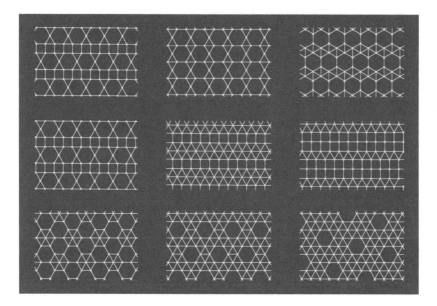

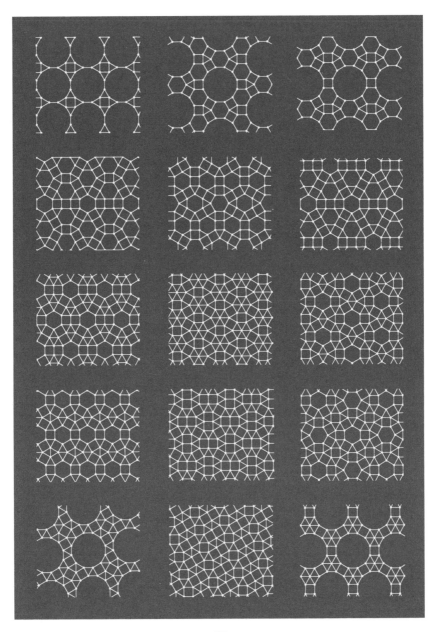

THE SMALLEST PART
reversible stencils and rotatable blocks

Many of the semi and demi-regular grids can be reduced to a simple square or triangular unit which can then be reflected or rotated to recreate the whole pattern. Often these repeat triangles or squares are surprisingly small. It is worth remembering, however, that in practical applications it is often easier to rotate a printing block or stencil than it is to reflect it—and in such cases one has to double the stencil or carve a larger block.

The design shown opposite is based on one of the grids on the previous page (see if you can find which one). It is produced by rotation *and* reflection of the primary unit (*below right*). Once this basic unit has been identified, all that you need to do is draw this minimum amount (*opposite top*) to be able to create the entire design.

Squares and equilateral triangles can both be halved to produce smaller triangular units (*below left*). But, again, take care and think about what you are doing; for instance you cannot do this with the example shown opposite—can you see why?

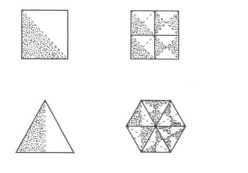

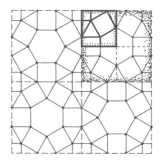

The smallest unit of the pattern below, drawn at the same scale, and showing the grid lines which have informed its design.

Notice how leaves and petals have been positioned along the edges of the tile, with the centers of flowers at the centers of rotation points.

Symmetry
regular and beautiful

Symmetry means 'measuring together', and things are termed symmetrical when they possess harmonious proportions, often between repeated elements. Elements may be repeated in a number of different ways: displaced, reflected, rotated, spiraled, scaled, stretched, folded, or multiple combinations of these.

Symmetries can be manifest (clear) or occult (hidden). For example, the balanced weights below (*center*) hint at the equations of mathematics and physics that model the hidden symmetries which underlie the physical world. Symmetry is the subject of an entire book in this series, and the images below and opposite are suggestive rather than complete. Apart from the symmetries listed above, there are also topological or mapping symmetries (*below right*), branching symmetries, fractal symmetries (where parts are images of the whole), crystal symmetries, electron orbital symmetries, aperiodic symmetries (*see Li symmetries, page 120*), radial symmetries, permutation symmetries, series symmetries (*for example phyllotaxis, see page 324*) and species of assymmetries.

It is important to realise that geometry and harmony are themselves merely forms of symmetry, especially in the way their products are appreciated as aesthetic. The plot of a film or novel can likewise be thought of as symmetrical, as can notions of fate or justice.

'Measuring together' can indeed mean many things.

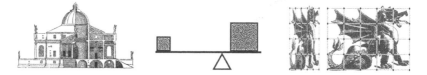

ISLAMIC DESIGNS

stars are born from subgrids

Islamic patterns speak of infinity and the omnipresent center.

For the pattern opposite, start with six circles round one, developing a grid of overlapping dodecagons from triangles, squares and hexagons (*see pages 73 and 99*). The key points now are halfway along the side of every polygon. These are joined up in a special way and extended as shown in the top part of the diagram. Many beautiful patterns are sitting in every simple subgrid, just waiting to be pulled out.

The subgrids themselves are rarely shown in traditional art. They are considered part of the underlying structure of reality, with the cosmos overlaid—'cosmos' means 'adornment'.

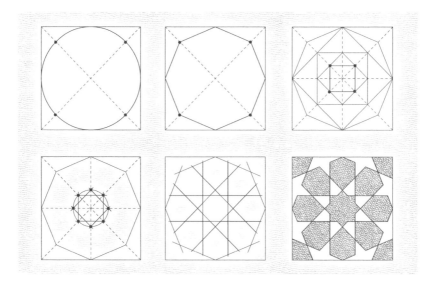

A Church Window

not far from the Isle of Man

A piece of church window masonry is shown opposite. The design speaks of the implicit trinity in unity. It is a very beautiful design and remarkably satisfying to draw. See if you can follow its construction from the diagram below, which begins with the enclosing circle. Notice how every detail is defined by the geometry.

Draw a large circle and divide it into six. In this circle draw a large triangle and inscribe a circle to fit inside it. This gives the centers of three touching circles (the center-lines of the tracery). Notice how these circles do not touch either the outer circle or the center of the window. A small circle (*at the bottom of the image below*) then assists in giving the width of the stone tracery itself, enabling the inner, middle and outer edges of tracery to be defined.

Now see if you could draw the tracery on the next page.

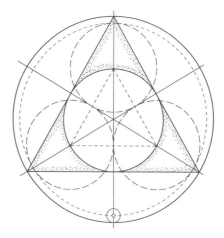

TREFOILS AND QUATREFOILS
the geometry behind tracery

Everything is made of light, all matter is, and without matter there would be no sound. Atoms and planets arrange themselves in geometrical patterns. How profound then is a window, which allows the passage of light into an otherwise dark space.

The designs of church windows follow many rules, forms and traditions, and some clues are given on these pages. The easiest to draw are the three quatrefoils (*bottom row of this page*).

The south window of Lincoln cathedral with its striking double *vesica* is shown opposite, and below it three famous and very early west windows, from Chartres, Evreux and Rheims cathedrals. A good balance is kept between line and curve.

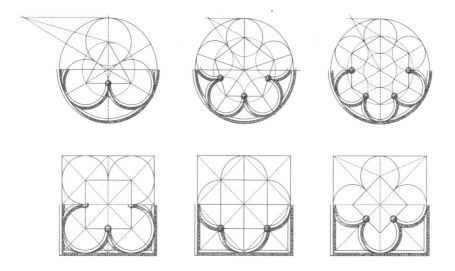

Lincoln

Chartres

Evreux

Rheims

III

STONE CIRCLES AND CHURCHES

vesicas in action for over 4,000 years

Four flattened stone circles are shown opposite with their consistent geometry as discovered by Professor Thom. On the left are examples of the type-A shape, on the right type-BS. The *vesica*-based constructions are also shown (*see page 68 for vesica*).

Shown on this page is the ground plan of Winchester Cathedral. An interplay of simple *vesica*-based triangular and square systems, *ad triangulum* and *ad quadratum*, underlies the plans of many ecclesiastical buildings (*see top row page 83*).

The design of a sacred building, whether church, stone circle or temple, requires the designer to marry the universal symbolism of the geometrical moves he or she is making with the specific religious language. Local factors also go into the cauldron, for example Sun, star or Moon rise and set positions, or nearby sacred hills, springs or leys. Below we see Winchester Cathedral's axis pointing 72° from north, creating a magical magnetic pentagram.

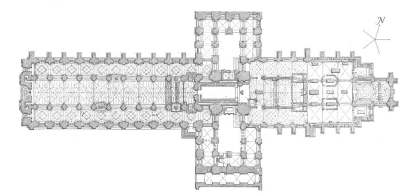

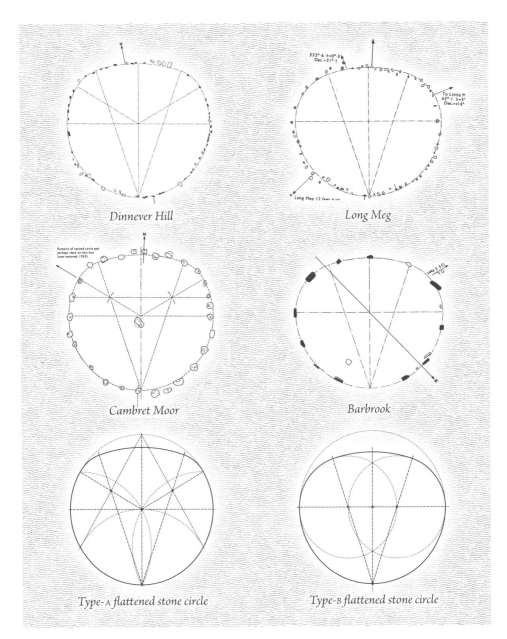

Dinnever Hill

Long Meg

Cambret Moor

Barbrook

Type-A flattened stone circle

Type-B flattened stone circle

DELIGHTFUL ARCHES
how to draw a few of the many

Arches take remarkably similar forms all over the world and a few are shown here. Living trees often make the best arches.

The top row opposite shows five two-centered arches. Their span has been divided into 2, 3, 4, 5, and 5 again. The straight dotted lines show the radii of their arcs. The heights of arches can vary but for these five their heights are defined by a rectangle which gives a musical interval, thus 2:3, 3:4 and so on (*page 84*).

The second row of arches opposite are four-centered. The curve of the arch changes at positions given by the solid line. Ideas for defining their heights are also given.

The bottom two arches opposite are a horseshoe arch, which can also be pointed, and a pointed arch. The pointed arch seems to turn up—'the return'—but the lines are actually dead straight.

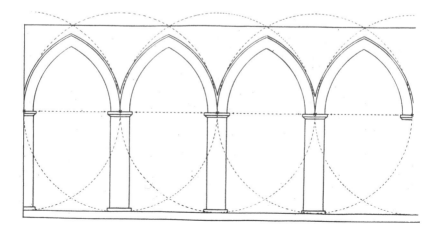

A CELTIC SPIRAL

Euclidean geometry in ancient Ireland

The design shown opposite comes from a four inch bronze disc found on Loughan Island in Northern Ireland, and dated to around 2000 years ago. It is an exceptionally beautiful example of the early Celtic style. As we have already seen with stone circles and arches, the seamless conjunction of multiple arcs can be highly aesthetic and it reached its perfection in the early Celtic period.

Many early Celtic pieces show evidence of compass use and the final drawing for this disc required no less than 42 separate compass-point positions! It is thought that the master artists who created these designs started with a basic geometric template, such as a touching circles pattern, then sketched their forms before returning to geometry to tighten everything up so that their curves all became arcs, sections of circles. This gives a tautness to the curves. In this way intuition and intellect work together.

The lower sequence of pictures shows how to plot arcs through points. The first diagram shows an arc centered on ‹c›. We want the arc to change effortlessly at ‹a› and then pass through ‹b›. What do we do? Find the perpendicular bisector between points ‹a› and ‹b› by opening the compass, describing two equal arcs from ‹a› and ‹b›, and drawing the line through their intersections (*lower, opposite center*). This cuts the ‹a–c› line at a new point ‹o› which then becomes the center we were looking for (*lower right*).

All of the beautiful curves in the Loughan Island disc are drawn and tautened in this simple and elegant manner.

The idea that circles or arcs of circles are the only truly heavenly curves finds its way into many philosophies of sacred art. The technique above may be used in any application to refine a curvy sketch or design and give it that elusive magical quality.

PENTAGONAL POSSIBILITIES
those phantastic phizzy phives

Although the regular pentagon does not tile on the plane, it does do various other things which no good book on sacred geometry should omit to mention. One of these, taken from studies by Johannes Kepler (1571–1630), is shown (*opposite top*), where a 'seed' pattern can be grown *from the center*. The grey pentagons leave spaces which are bits of pentagrams and vice versa. The design is riddled with examples of the Golden Section. Other 'seeds' are shown.

The mathematician Roger Penrose recently discovered the tiling shown lower, opposite. Two shapes tile to fill the plane with aperiodic 'not-repeating' pentagonal elements at all scales. These patterns have recently been found to underlie the nature of most liquids. They are, for instance, cross-sections through water.

Shown below are the fourteen types of irregular convex pentagon which tile the plane.

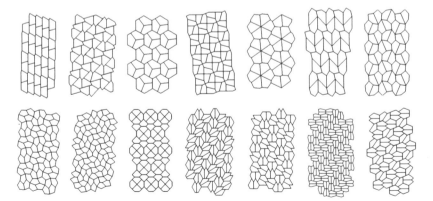

*Kepler's early 17th century infinite pentagonal system and
below, Penrose's 20th century pair of tiles which fill the plane*

LI SYMMETRIES
formed in time

Li symmetries are so familiar to us that we almost don't notice them. They surround us and pervade the natural world, but it was only in the 1950s that these enigmatic forms of symmetry began to be understood as self-organising systems through the pioneering work of Alan Turing. The Chinese, however, have been studying them for millennia, and it is from them that they get their name.

Li symmetries may be distinguished from static symmetries in that they are primarily caused by the interaction between processes and materials. For instance the repeated action of wind over sand produces the familiar ribbing of sand dunes, a symmetry which can occur at different scales. Likewise, the action of heat on wet clay creates crack patterns which also closely ressemble the layout of many towns and cities, even down to relatively small details such as the width of the roads (*see David Wade's illustrations opposite*).

Li symmetries extend into animal markings, stretch patterns such as tree barks, cloud patterns, and many other areas of nature.

Next time you are out and about, see how many you can spot!

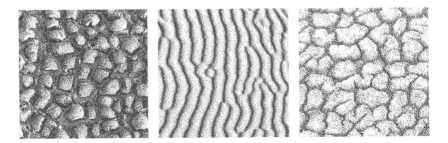

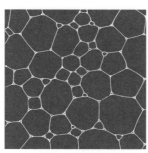
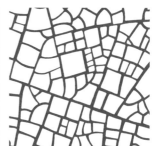

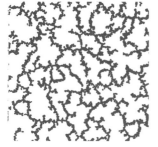

THE SEVENTEEN SYMMETRIES
from slide, spin and mirror

The Arab alchemist Jabir ibn Hayyan, known in the West as Geber, regarded 17 as the numerical basis of the physical world.

Using a very simple sample design the next three pages explore the three basic operations of rotation, reflection and sliding. These, combined with the three regular tilings, give seventeen 'wallpaper' or plane symmetry groups which are shown below, opposite, and on the next page (*after Critchlow*). The final facing page shows the seven possible frieze symmetries derived in the same way.

This visual key can be very useful when creating repeats for fabric or pottery patterns (*see too pages 102–103*). 'Pattern', by the way, comes from the Latin word *pater,* meaning 'father', in the same way that 'matrix' comes from *mater,* meaning 'mother'.

Remember, not all stencils can be turned over (reflected) without making a mess, so choose your repeat units with care.

And on that rather practical note this dense little book on one of the oldest subjects on Earth has now reached its end. I hope you have gleaned enough ideas from it to create something good, true and beautiful next time you get designing!

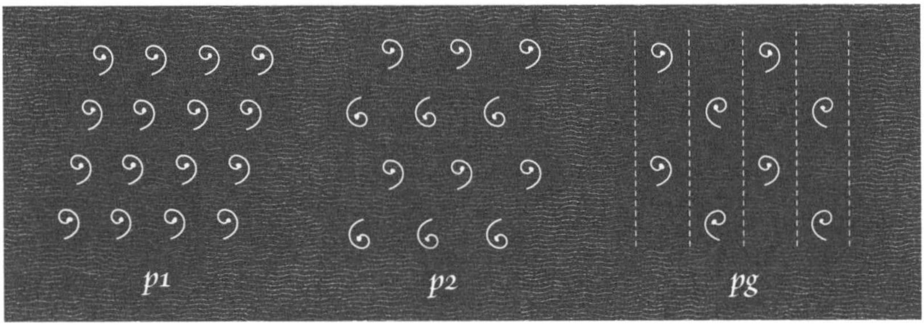

p1　　　　　　　p2　　　　　　　pg

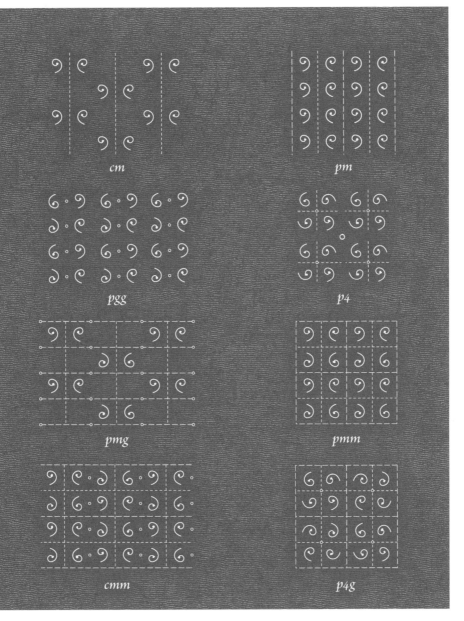

cm

pm

pgg

p4

pmg

pmm

cmm

p4g

p4m

p3

p3m1

p6

p31m

p6m

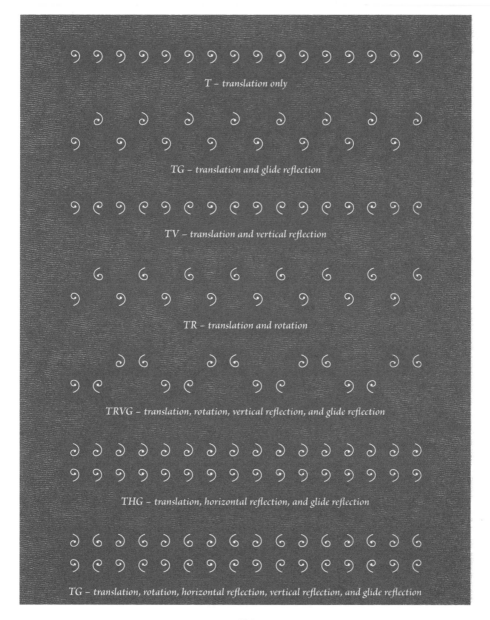

T – translation only

TG – translation and glide reflection

TV – translation and vertical reflection

TR – translation and rotation

TRVG – translation, rotation, vertical reflection, and glide reflection

THG – translation, horizontal reflection, and glide reflection

TG – translation, rotation, horizontal reflection, vertical reflection, and glide reflection

BOOK III

PLATONIC
& ARCHIMEDEAN
SOLIDS

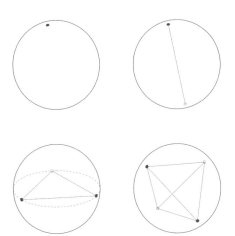

Daud Sutton

INTRODUCTION

Imagine a sphere. It is Unity's perfect symbol. Each point on its surface is identical to every other, equidistant from the unique point at its center.

Establishing a single point on the sphere allows others to be defined in relation to it. The simplest and most obvious relationship is with the point directly opposite, found by extending a line through the sphere's center to the other side. Add a third point and space all three as far from each other as possible to define an equilateral triangle. The three points lie on a circle with radius equal to the sphere's and sharing its center, an example of the largest circles possible on a sphere, known as *great circles*. Point, line and triangle occupy zero, one and two dimensions respectively. It takes a minimum of four points to define an uncurved three-dimensional form.

This section of *Quadrivium* charts the unfolding of number in three-dimensional space through the most fundamental forms derived from the sphere. A cornerstone of mathematical and artistic inquiry since antiquity, after countless generations these beautiful forms continue to intrigue and inspire.

THE PLATONIC SOLIDS
beautiful forms unfold from unity

Imagine you are on a desert island; there are sticks, stones, and sheets of bark. If you start experimenting with three-dimensional structures you may well discover five 'perfect' shapes. In each case they look the same from any *vertex* (corner point), their faces are all made of the same regular shape, and every edge is identical. Their vertices are the most symmetrical distributions of four, six, eight, twelve, and twenty points on a sphere (*below*).

These forms are examples of *polyhedra*, literally 'many seats' and, as the earliest surviving description of them as a group is in Plato's *Timaeus*, they are often called the Platonic Solids. Plato lived from 427 to 347 BC, but there is evidence that the Platonic Solids were discovered much earlier.

Three of the solids have faces of equilateral triangles—three, four, or five meeting at each vertex—and have names deriving from their number of faces; the *tetrahedron* is made from four, the *octahedron* eight, and the *icosahedron* twenty. The 3-4-5 theme continues with the common *cube*, with its six square faces, and the *dodecahedron* with its twelve regular pentagonal faces. Over the pages which follow we will get to know these striking three-dimensional forms better.

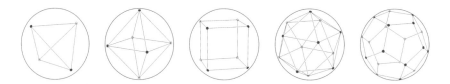

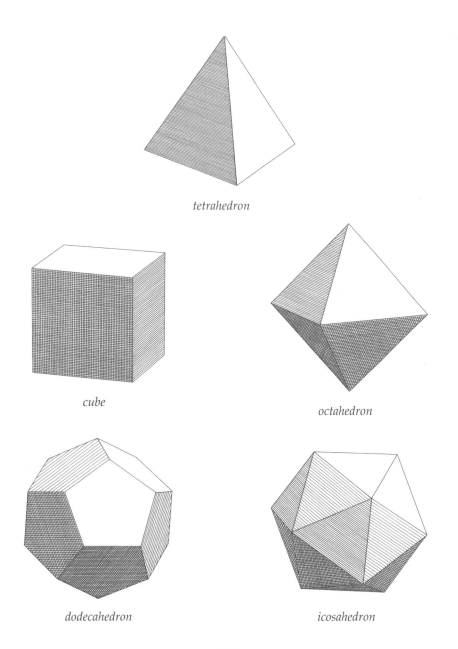

tetrahedron

cube

octahedron

dodecahedron

icosahedron

THE TETRAHEDRON
4 faces : 6 edges : 4 vertices

The tetrahedron is composed of four equilateral triangles, with three meeting at every vertex. Its vertices can also be defined by the centers of four touching spheres (*lower, opposite right*). Plato associated its form with the element of Fire because of the penetrating acuteness of its edges and vertices, and because it is the simplest and most fundamental of the regular solids. The Greeks also knew the tetrahedron as *puramis*, whence the word *pyramid*. Curiously the Greek word for fire is *pur*.

The tetrahedron has three 2-fold axes, passing through the midpoints of its edges, and four 3-fold axes, each passing through one vertex and the center of the opposite face (*below*). Any polyhedron with these axes of rotation has *tetrahedral symmetry*.

Each Platonic Solid is contained by its *circumsphere*, which just touches every vertex. The Solids also define two more spheres: their *midsphere*, which passes through the midpoint of every edge, and their *insphere*, which is contained by the solid, perfectly touching the center of every face. For the tetrahedron the *inradius* is one third of the *circumradius* (*lower, opposite left*).

edge on : 2-fold

face on : 3-fold

from vertex : 3-fold

THE OCTAHEDRON
8 faces : 12 edges : 6 vertices

The octahedron is made of eight equilateral triangles, four meeting at every vertex. Plato considered the octahedron an intermediary between the tetrahedron, or Fire, and the icosahedron, or Water, and thus ascribed it to the element of Air. The octahedron has six 2-fold axes passing through opposite edges, four 3-fold axes through its face centers and three 4-fold axes through opposite vertices (*below*). Solids combining these rotation axes display *octahedral symmetry*.

Greek writings attribute the discovery of the octahedron and icosahedron to Theaetetus of Athens (417–369 BC). Book XIII of Euclid's *Elements* (*see page 144*) is thought to be based on Theaetetus' work on the regular solids.

The octahedron's circumradius is bigger than its inradius by a factor of $\sqrt{3}$ (*see page 377*). The same relationship occurs between the circumradius and inradius of the cube, and between the circumradius and *midradius* (and the midradius and inradius) of the tetrahedron.

The tetrahedron, the octahedron and the cube are all found in the mineral kingdom. Mineral diamonds and common fluorite crystals often form octahedra.

edge on : 2-fold

face on : 3-fold

from vertex : 4-fold

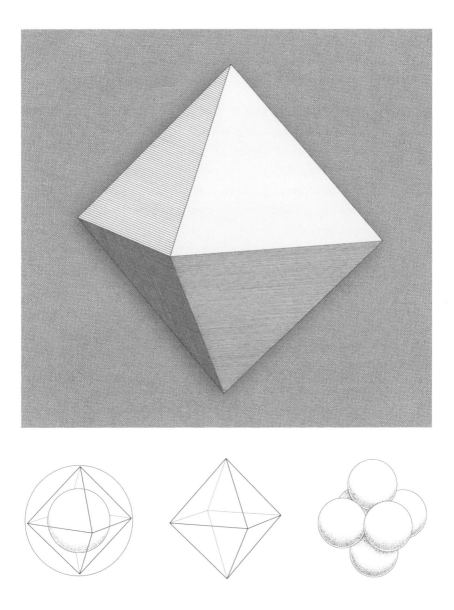

THE ICOSAHEDRON

20 faces : 30 edges : 12 vertices

The icosahedron is composed of twenty equilateral triangles, five to a vertex. It has fifteen 2-fold axes, ten 3-fold axes and six 5-fold axes (*below*), known as *icosahedral symmetry*. When the tetrahedron, octahedron, and icosahedron are made of identical triangles, the icosahedron is the largest. This led Plato to associate the icosahedron with Water, the densest and least penetrating of the three fluid elements: Fire, Air, and Water.

The angle where two faces of a polyhedron meet at an edge is known as a *dihedral angle*. The icosahedron is the Platonic Solid with the largest dihedral angles.

If you join the two ends of an icosahedron's edge to the center of the solid an isosceles triangle is defined—the same as the triangles that make up the faces of the Great Pyramid at Giza. Opposing edges of an icosahedron form Golden Section rectangles (*see page 152*).

Arranging twelve equal spheres to define an icosahedron leaves space at the center for another sphere just over nine tenths as wide as the others (*lower, opposite right*).

edge on : 2-fold

face on : 3-fold

from vertex : 5-fold

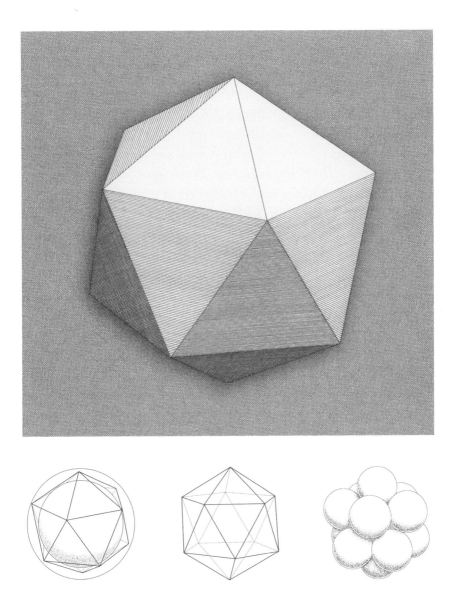

THE CUBE

6 faces : 12 edges : 8 vertices

The cube has octahedral symmetry (*below*). Plato assigned it to the element of Earth due to the stability of its square bases. Aligned to our experience of space, it faces forwards, backwards, right, left, up, and down, corresponding to the six directions North, South, East, West, zenith, and nadir. As we saw in BOOK I of this volume, six is the first perfect number, with factors adding up to itself $(1 + 2 + 3 = 6)$.

Add the cube's twelve edges, the twelve face diagonals and the four interior diagonals to find a total of twenty-eight straight paths joining the cube's eight vertices to each other. Twenty-eight is the second perfect number $(1 + 2 + 4 + 7 + 14 = 28)$.

Islam's annual pilgrimage is to the Kaaba, literally Cube, in Mecca. The sanctuary of the Temple of Solomon was a cube, as is the crystalline New Jerusalem in Saint John's revelation. In 430 BC the oracle at Delphi instructed the Athenians to double the volume of the cubic altar of Apollo, whilst maintaining its shape. 'Doubling the cube', as the problem became known, ultimately proved impossible using Euclidean geometry alone.

edge on : 2-fold

from vertex : 3-fold

face on : 4-fold

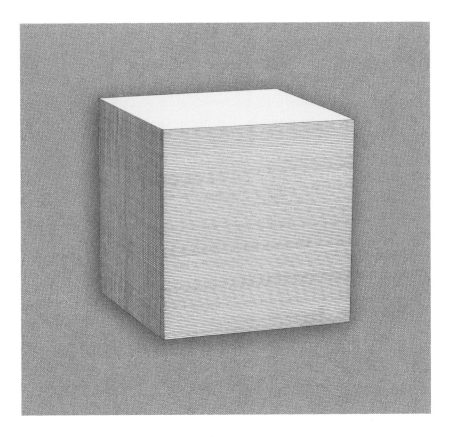

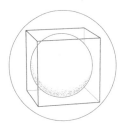 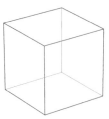 

The Dodecahedron
12 faces : 30 edges : 20 vertices

The beautiful dodecahedron has twelve regular pentagonal faces, three of which meet at every vertex. Its symmetry is icosahedral (*below*). Like the tetrahedron, or pyramid, and the cube, the dodecahedron was known to the early Pythagoreans and was commonly referred to as *the sphere of twelve pentagons*. Having detailed the other four solids and ascribed them to the elements, Plato's Timaeus says enigmatically "There remained a fifth construction which God used for embroidering the constellations on the whole heaven."

A dodecahedron sitting on a horizontal surface has vertices lying in four horizontal planes which cut the dodecahedron into three parts. Surprisingly, the middle part is equal in volume to the others, so each is one third of the total! Also, when set in the same sphere, the surface areas of the icosahedron and dodecahedron are in the same ratio as their volumes, and their inspheres are identical.

'Fool's Gold', or *iron pyrite*, forms crystals much like the dodecahedron, but don't be fooled, their pentagonal faces are not regular and their symmetry is tetrahedral.

edge on : 2-fold   *from vertex : 3-fold*   *face on : 5-fold*

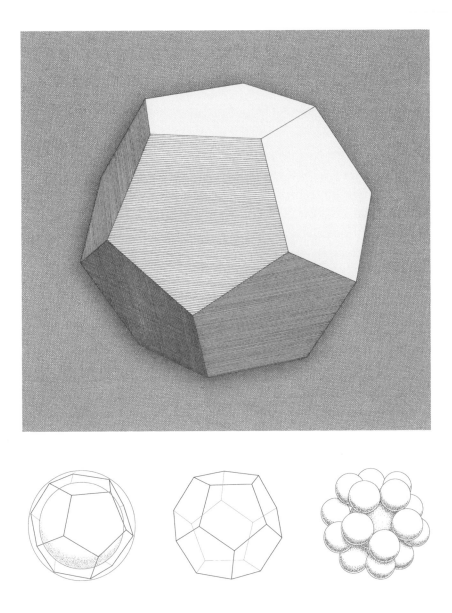

A SHORT PROOF
are there really only five?

A *regular polygon* has equal sides and angles. A *regular polyhedron* has equal regular polygon faces and identical vertices. The Platonic Solids are the only possible *convex regular polyhedra*. In book XIII of his *Elements* Euclid of Alexandria (*ca.* 325–265 BC) proves that each of these five convex regular polyhedra can be constructed, and concludes by demonstrating that there are no other possibilities.

At least three polygons are needed to make a *solid angle*. Using equilateral triangles this is possible with three ‹*a*›, four ‹*b*› and five ‹*c*› around a point. With six the result lies flat ‹*d*›. Three squares make a solid angle ‹*e*›, but with four ‹*f*› a limit similar to six triangles is reached. Three regular pentagons form a solid angle ‹*g*›, but there is no room, even lying flat, for four or more. Three regular hexagons meeting at a point lie flat ‹*h*›, and higher polygons cannot meet with three around a point, so a final limit is reached. Since only five solid angles made of identical regular polygons are possible, there are at most five possible convex regular polyhedra.

The angle left as a gap when a polyhedron's vertex is folded flat is its *angle deficiency*. René Descartes (1596–1650) discovered that the sum of a convex polyhedron's angle deficiencies always equals 720°, or two full turns. Later, in the eighteenth century, Leonhard Euler (1707–1783) noticed another peculiar fact: in every convex polyhedron the number of faces minus the number of edges plus the number of vertices equals two.

ALL THINGS IN PAIRS
platonic solids two by two

What happens if we join the face-centers of the Platonic Solids? Starting with a tetrahedron, we discover another, inverted, tetrahedron. The face-centers of a cube produce an octahedron, and an octahedron creates a cube. The icosahedron and dodecahedron likewise produce each other. Two polyhedra whose faces and vertices correspond perfectly are known as each other's *duals*. The tetrahedron is *self-dual*. Dual polyhedra have the same number of edges and the same symmetries.

The illustrations opposite are stereogram pairs. Hold the book at arm's length and place a finger vertically, midway to the page. Focus on the finger and then bring the central blurred image into focus. The image should jump into three dimensions!

Dual pairs of Platonic Solids can be married with their edges touching at their midpoints to give the compound polyhedra shown below. Everything in Creation has its counterpart or opposite, and the dual relationships of the Platonic Solids are a beautiful example of this principle.

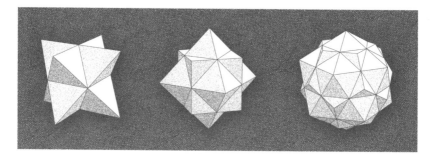

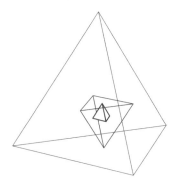
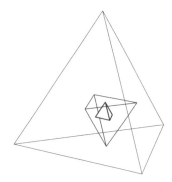

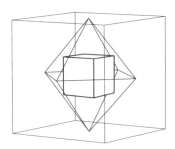
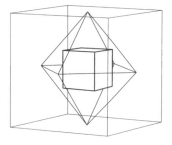

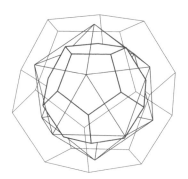
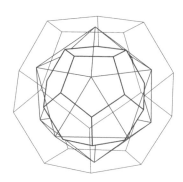

AROUND THE GLOBE
in elegant ways

Plato's cosmology constructs the Elemental Solids from two types of right-triangular atom. The first atom is half an equilateral triangle, six of which then compound to produce larger equilateral triangles; these go on to form the tetrahedron, octahedron and icosahedron. The second triangular atom is a diagonally halved square, which appears in fours, making squares which then form cubes.

The Platonic Solids have planes of symmetry dividing them into mirror image halves; the tetrahedron has six mirror planes, the octahedron and cube have nine, and the icosahedron and dodecahedron have fifteen. When the tetrahedron, octahedron and icosahedron are constructed from Plato's triangular atoms, paths are defined which make their mirror planes explicit. The cube however needs twice as many triangular divisions as Plato gave it (*top row*) to delineate all its mirror planes (*middle row*).

Projecting the subdivided Platonic Solids onto their circumspheres produces three spherical systems of symmetry. Each spherical system is defined by a characteristic spherical triangle with one right angle, and one angle of one third of a half turn. Their third angles are respectively one third of a half turn (*top row*), one quarter of a half turn (*middle row*) and one fifth of a half turn (*lower row*). This sequence of ⅓, ¼, and ⅕ elegantly inverts the Pythagorean whole number triple 3, 4, 5.

ROUND AND ROUND
lesser circles

Any navigator will tell you that the shortest distance between two points on a sphere's surface is always an arc of a *great circle*. When a polyhedron's edges are projected onto its circumsphere the result is a set of great circle arcs known as a *radial projection*. Opposite, the left hand column shows the radial projections of the Platonic Solids with their great circles shown in dotted line.

A spherical circle smaller than a great circle is called a *lesser circle*. Tracing a circle around all the faces of the Platonic Solids set in their circumspheres generates the patterns of lesser circles shown in the middle column. The apocryphal book XIV of Euclid's *Elements* proves that when set in the same sphere, the lesser circles around the dodecahedron's faces (*fourth row*) are equal to the lesser circles around the icosahedron's faces (*fifth row*). The same is true of the cube (*second row*) and the octahedron (*third row*) as a pair.

Shrink the lesser circles in the middle column until they just touch each other to define the five spherical curiosities in the right hand column. Many neolithic carved stone spheres have been found in Scotland carved with the same patterns as the first four of these arrangements (*see page 74*). The dodecahedral carvings of twelve circles on a sphere, some 4,000 years old, are the earliest known examples of man-made designs with icosahedral symmetry.

Large lesser circle models can be made from circles of willow, or cheap hula-hoops, lashed together with wire, string, or tape.

THE GOLDEN SECTION
and some intriguing juxtapositions

Dividing a line so that the shorter section is to the longer as the longer section is to the whole line defines the Golden Section (*below*). The Golden Section proportion is an irrational number, inexpressible as a simple fraction (*see pages 54 and 377*). Its value is one plus the square root of five, divided by two—approximately 1.618. It is represented by the Greek letter Φ (*phi*), or sometimes by τ (*tau*). Φ has intimate connections with unity; Φ times itself (Φ²) is equal to Φ plus one (2.618...), and one divided by Φ equals Φ minus one (0.618...). It is innately related to fivefold symmetry; the heavy lines in the pentagram below form a continuous series of Golden Section relationships.

Remove a square from one side of a Golden Section rectangle and the remaining rectangle will also have sides in the Golden Section. This process can continue indefinitely and establishes a Golden Section spiral (*below right*). Remarkably, an icosahedron's twelve vertices are defined by three perpendicular Golden Section rectangles (*opposite top*). The dodecahedron is richer still. Twelve of its twenty vertices are defined by three perpendicular Φ² rectangles, and the remaining eight vertices are found by adding a cube of edge length Φ (*lower, opposite*).

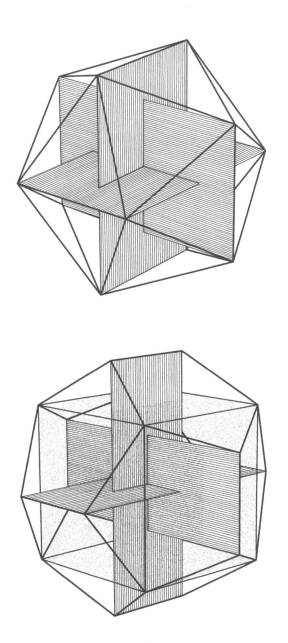

POLYHEDRA WITHIN POLYHEDRA
and so proceed ad infinitum

The Platonic Solids fit together in remarkable and fascinating ways; the appendix on page 376 shows many relationships. The upper stereogram pair opposite shows a dodecahedron with edge length one. Nested inside it is a cube, edge length Φ, and a tetrahedron, edge length √2 times the cube's (*see page 377*). The tetrahedron occupies one third of the cube's volume.

In the lower stereogram pair opposite, the six edge midpoints of the tetrahedron define the six vertices of an octahedron. As well as halving the tetrahedron's edges this octahedron has half its surface area and half its volume, perfectly embodying the musical octave ratio of 1:2. Similarly the twelve edges of the octahedron correspond to the twelve vertices of a nested icosahedron. The icosahedron's vertices cut the octahedron's edges perfectly in the Golden Section (*see page 146 for instructions on how to make these stereogram images to jump into 3-D*).

Imagine these two sets of nestings combining to give all five Platonic Solids in one elegant arrangement. Since the outer dodecahedron defines a larger icosahedron by their dual relationship, and the inner icosahedron likewise defines a smaller dodecahedron, the nestings can be continued outwards and inwards to infinity.

The tetrahedron, octahedron, and icosahedron, made entirely from equilateral triangles, are *convex deltahedra*. There are only five other possible convex deltahedra, all shown lower, opposite.

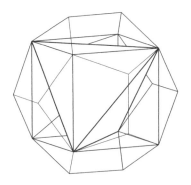
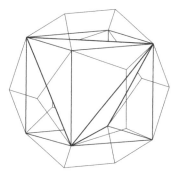

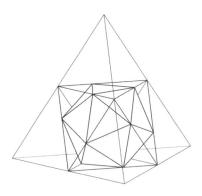
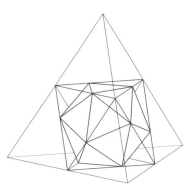

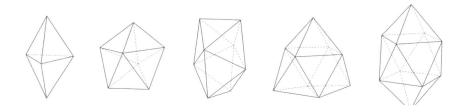

COMPOUND POLYHEDRA
a stretch of the imagination

The interrelationships on the previous page generate particularly beautiful compound polyhedra. Fix the position of an icosahedron, and octahedra can be placed around it in five different ways, giving the compound of five octahedra (*top left*). Similarly the cube within the dodecahedron, placed five different ways, generates the compound of five cubes (*top right*). The tetrahedron can be placed in the cube two different ways to give the compound of two tetrahedra shown on page 146. Replace each of the five cubes in the dodecahedron with two tetrahedra to give the compound of ten tetrahedra (*middle left*). Remove five of the tetrahedra from the compound of ten, to leave the compound of five tetrahedra (*middle right*). This occurs in two versions, right-handed or *dextro* and left-handed or *laevo*; the two versions cannot be superimposed and are described as each others' *enantiomorphs*. Polyhedra or compounds with this property of 'handedness' are described as *chiral*.

Returning to the cube and dodecahedron, and this time fixing the cube, there are two ways to place the dodecahedron around it. The result of both ways used simultaneously is the compound of two dodecahedra (*lower left*). In the same way the octahedron and icosahedron pair gives the compound of two icosahedra (*lower right*). Many other extraordinary compound polyhedra are possible, for example Bakos' compound of four cubes is shown on page 128.

THE KEPLER POLYHEDRA
the stellated and great stellated dodecahedron

The sides of some polygons can be extended until they meet again, for example the regular pentagon extends to form a five-pointed star, or pentagram (*below*). This process is known as *stellation*. Kepler had a great fascination with polyhedra (*see for example page 306*) and proposed the application of stellation to them, observing the two possibilities of stellation by extending edges, and stellation by extending face planes. Applying the first of these to the dodecahedron and icosahedron (*below*) he discovered the two polyhedra illustrated opposite and named them the larger and smaller icosahedral hedgehogs!

Their modern names, the stellated dodecahedron (*opposite top*) and the great stellated dodecahedron (*lower, opposite*), reveal that these polyhedra are also two of the face stellations of the dodecahedron. Each is made of twelve pentagram faces, one with five, the other with three to every vertex. They have icosahedral symmetry.

Although its five sides intersect each other, the pentagram has equal edges and equal angles at its vertices and so can be considered a *non-convex regular polygon*. Likewise, these polyhedra can be regarded as *non-convex regular polyhedra*.

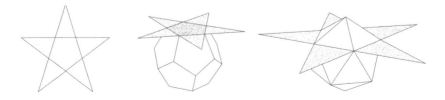

THE POINSOT POLYHEDRA
the great dodecahedron and great icosahedron

Louis Poinsot (1777–1859) investigated polyhedra independently of Kepler, rediscovering Kepler's two icosahedral hedgehogs and also discovering the two polyhedra shown here, the great dodecahedron (*top*) and the great icosahedron (*lower*). Both of these polyhedra have five faces to a vertex, intersecting each other to give pentagram *vertex figures*. The great dodecahedron has twelve pentagonal faces and is the third stellation of the dodecahedron. The great icosahedron has twenty triangular faces and is one of an incredible fifty-eight possible stellations of the icosahedron (often numbered as fifty-nine including the icosahedron itself). These stellations also include the compounds of five octahedra, five tetrahedra, and ten tetrahedra.

A non-convex regular polyhedron must have vertices arranged like one of the Platonic Solids. Joining a polyhedron's vertices to form new types of polygon within it is known as *faceting*. The possibilities of faceting the Platonic Solids produce the compounds of two and ten tetrahedra, the compound of five cubes, the two Poinsot polyhedra (*below left*) and the two Kepler star polyhedra (*below right*). The four Kepler-Poinsot polyhedra are therefore the only non-convex regular polyhedra.

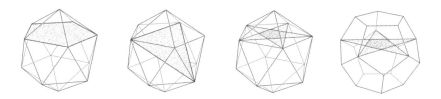

THE ARCHIMEDEAN SOLIDS
thirteen semi-regular polyhedra

The thirteen Archimedean Solids (*opposite*) are the subject of much of the rest of this book. Also known as the *semi-regular polyhedra*, they have regular faces of more than one type, and identical vertices. They all fit perfectly within a sphere, with tetrahedral, octahedral or icosahedral symmetry. Although their earliest attribution is to Archimedes, Kepler seems to have been the first since antiquity to describe the whole set of thirteen in his *Harmonices Mundi*. He further noted the two infinite sets of regular prisms and antiprisms (*examples below*) which also have identical vertices and regular faces.

Turn one octagonal cap of the rhombicuboctahedron by an eighth of a turn to obtain the pseudo-rhombicuboctahedron (*below*). Its vertices, while surrounded by the same regular polygons, are of *two* types relative to the polyhedron as a whole.

There are fifty-three semi-regular non-convex polyhedra, one example being the dodecadodecahedron (*below*). Together with the Platonic and Archimedean Solids, and the Kepler-Poinsot Polyhedra, they form the set of seventy-five *Uniform Polyhedra*.

heptagonal prism   *heptagonal antiprism*   *pseudo rhombicuboctahedron*   *dodecadodecahedron*

truncated tetrahedron

truncated octahedron

cuboctahedron

truncated cube

rhombicuboctahedron

great rhombicuboctahedron

snub cube

truncated icosahedron

icosidodecahedron

truncated dodecahedron

rhombicosidodecahedron

great rhombicosidodecahedron

snub dodecahedron

Five Truncations

off with their corners!

Truncate the Platonic Solids to produce the five equal-edged Archimedean polyhedra shown here. These truncated solids are the perfect demonstration of the Platonic Solids' vertex figures: triangular for the tetrahedron, cube and dodecahedron, square for the octahedron and pentagonal for the icosahedron. Each Archimedean Solid has one circumsphere and one midsphere. They have an insphere for each type of face, the larger faces having the smaller inspheres touching their centers. Each truncated solid therefore defines four concentric spheres.

The five truncated solids can each sit neatly inside both their original Platonic Solid and that Platonic Solid's dual. For example the truncated cube can rest its octagonal faces within a cube or its triangular faces within an octahedron.

The truncated octahedron is the only Archimedean Solid that can fill space with identical copies of itself, leaving no gaps. It also conceals a less obvious secret. Joining the ends of one of its edges to its center produces a central angle which is the same as the acute angle in the famous Pythagorean 3:4:5 triangle, beloved of ancient Egyptian masons for defining a right angle.

THE CUBOCTAHEDRON

14 faces : 24 edges : 12 vertices

The cuboctahedron combines the six square faces of the cube with the eight triangular faces of the octahedron. It has octahedral symmetry. Joining the edge midpoints of either the cube or the octahedron traces out a cuboctahedron (shown below as a stereogram pair). According to Heron of Alexandria (10–75 AD), Archimedes ascribed the cuboctahedron to Plato.

Quasiregular polyhedra such as the cuboctahedron are made of two types of regular polygon, each type being surrounded by polygons of the other type. The identical edges, in addition to defining the faces themselves, also define equatorial polygons. For example the cuboctahedron's edges define four regular hexagons (*lower, opposite center*). The radial projections of quasiregular polyhedra consist entirely of complete great circles (*lower, opposite left*).

Twelve spheres pack around an identical thirteenth to produce a cuboctahedron (*lower, opposite right*). Greengrocers use this system to stack oranges in offset hexagonal layers. Known to chemists as hexagonal close packing each sphere is surrounded by twelve others, their centers defining a strong lattice of tetrahedra and octahedra.

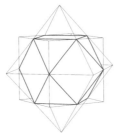 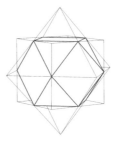

A Cunning Twist

and a structural wonder

Picture a cuboctahedron made of rigid struts joined at flexible vertices. This structure was named the *jitterbug* by R. Buckminster Fuller (1895–1983), and is shown opposite with the rigid triangular faces filled in for clarity. The jitterbug can be slowly collapsed in on itself in two ways so that the square 'holes' become distorted. When the distance between the closing corners equals the edge length of the triangles, an icosahedron is defined. Continue collapsing the structure and it becomes an octahedron. If the top triangle is then given a twist, the structure flattens to form four triangles which close up to give the tetrahedron.

Geodesic domes are another of Buckminster Fuller's structural discoveries. These are parts of geodesic spheres, which are formed by subdividing the faces of a triangular polyhedron, usually the icosahedron, into smaller triangles, and then projecting the new vertices outward to the same distance from the center as the original ones (*below*). A distant relative of the geodesic sphere is the popular Renaissance polyhedron of seventy-two sides known as Campanus' sphere (*below right*).

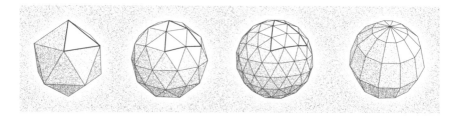

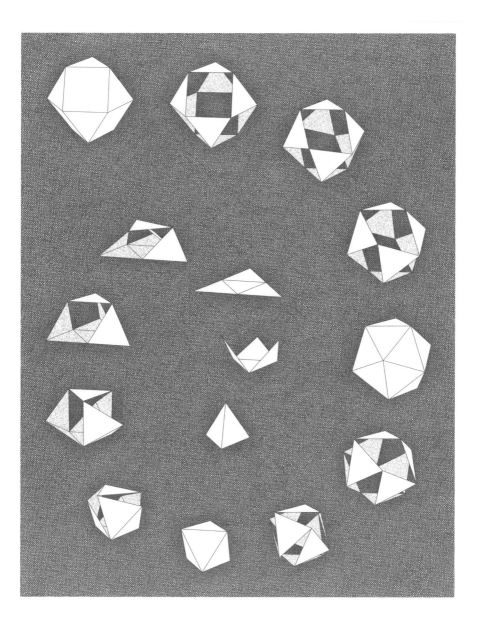

THE ICOSIDODECAHEDRON

32 faces : 60 edges : 30 vertices

The icosidodecahedron combines the twelve pentagonal faces of the dodecahedron with the twenty triangular faces of the icosahedron. Joining the edge midpoints of either the dodecahedron or the icosahedron traces out the quasiregular icosidodecahedron (*both are shown below as a stereogram pair*). Its edges form six equatorial decagons, giving a radial projection of six great circles (*lower, opposite left*).

The earliest known depiction of the icosidodecahedron is by Leonardo Da Vinci (1452–1519) and appears in Fra Luca Pacioli's (1445–1517) *De Divina Proportione*. Appropriately this work's main theme is the Golden Section, which is perfectly embodied by the ratio of the icosidodecahedron's edge to its circumradius.

Defining the icosidodecahedron with thirty equal spheres leaves space for a large central sphere that is $\sqrt{5}$ (*see page 377*) times as big as the others (*lower, opposite right*).

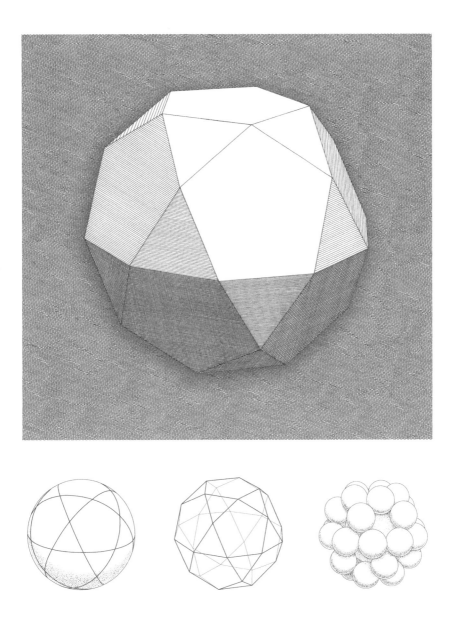

Four Explosions
expanding from the center

Exploding the faces of the cube or the octahedron outwards until they are separated by an edge length (*below*) defines the rhombicuboctahedron (*opposite top left*). The same process applied to the dodecahedron or icosahedron gives the rhombicosidodecahedron (*opposite top right*). The octagonal faces of the truncated cube, or the hexagonal faces of the truncated octahedron, explode to give the great rhombicuboctahedron (*lower, opposite left*). The decagonal faces of the truncated dodecahedron, or the hexagonal faces of the truncated icosahedron, explode to give the great rhombicosidodecahedron (*lower, opposite right*).

Kepler called the great rhombicuboctahedron a truncated cuboctahedron, and the great rhombicosidodecahedron a truncated icosidodecahedron. The two truncations he refers to, however, do not produce square faces, but √2 and Φ rectangles respectively.

These four polyhedra have face planes in common with either the cube, octahedron and rhombic dodecahedron (*see page 179*), or the icosahedron, dodecahedron, and rhombic triacontahedron (*see page 179*), hence the prefix 'rhombi-' in their names.

TURNING

the snub cube and snub dodecahedron

The name 'snub cube' is a loose translation of Kepler's name *cubus simus*, literally 'the squashed cube'. Both the snub cube and the snub dodecahedron are *chiral*, occurring in *dextro* and *laevo* versions. Both versions are illustrated opposite with the dextro versions on the right. The snub cube has octahedral symmetry, and the snub dodecahedron has icosahedral symmetry. Neither has any mirror planes. Of the Platonic and Archimedean Solids the snub dodecahedron is closest to the sphere.

The rhombicuboctahedron (*previous page*) can be used to make a structure similar to the jitterbug (*see page 169*). Applying a twist to this new structure produces the snub cube (*below*). Twist one way to make the dextro version and the other to make the laevo. The corresponding relationship exists between the rhombicosidodecahedron and the snub dodecahedron.

The five Platonic Solids have been truncated, combined, exploded and twisted into the thirteen Archimedean Solids. Three-dimensional space is revealing its order, complexity and subtlety. What other wonders await?

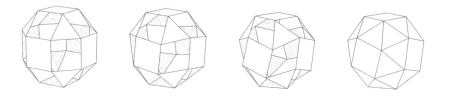

THE ARCHIMEDEAN DUALS

everything has its opposite

The duals of the Archimedean Solids were first described as a group by Eugène Catalan (1814–1894) and are positioned opposite to correspond with their partners on page 163. To create the dual of an Archimedean Solid, extend perpendicular lines from its edge midpoints, tangential to the Solid's midsphere. These lines are the dual's edges; the points where they first intersect each other are its vertices. Archimedean Solids have one type of vertex and different types of faces, their duals therefore have one type of face but different types of vertices.

The two quasiregular Archimedean Solids, the cuboctahedron and the icosidodecahedron, both have rhombic duals which were discovered by Kepler. The Platonic dual pair compounds (*pages 146, 166, and 170*) define the face diagonals of these rhombic polyhedra, which are in the ratios √2 for the rhombic dodecahedron and Φ for the rhombic triacontahedron. Kepler noticed that bees terminate their hexagonal honeycomb cells with three such √2 rhombs. He also described the three dual pairs involving quasiregular solids (*below*), where the cube is seen as a rhombic solid, and the octahedron as a quasiregular solid.

triakistetrahedron

tetrakishexahedron

rhombic dodecahedron

triakisoctahedron

trapezoidal icositetrahedron

disdyakisdodecahedron

pentagonal icositetrahedron

pentakisdodecahedron

rhombic triacontahedron

triakisicosahedron

trapezoidal hexecontahedron

disdyakistriacontahedron

pentagonal hexecontahedron

177

More Explosions
and unseen dimensions

Exploding the rhombic dodecahedron, or its dual the cuboctahedron, results in an equal-edged convex polyhedron of fifty faces (*opposite top right*). Meanwhile, the exploded rhombic triacontahedron, which is identical with an exploded icosidodecahedron, has one hundred and twenty-two faces (*lower, opposite right*).

Ludwig Schläfi (1814–1895) proved that there are six regular four-dimensional *polytopes* (generalisations of polyhedra): the 5-cell made of tetrahedra, the 8-cell or *tesseract* made of cubes, the 16-cell made of tetrahedra, the 24-cell made of octahedra, the 120-cell made of dodecahedra, and the 600-cell made of tetrahedra.

The rhombic dodecahedron is a three-dimensional shadow of the four-dimensional tesseract analogous to the hexagon as a two-dimensional shadow of the cube. In a cube two squares meet at every edge. In a tesseract three squares meet at every edge. Squares through the same edge define three cubes (*shaded below, with an alternative tesseract projection*).

Schläfi also proved that in five or more dimensions the only regular polytopes are the *simplex*, or generalized tetrahedron, the *hypercube*, or generalized cube, and the *orthoplex*, or generalized octahedron.

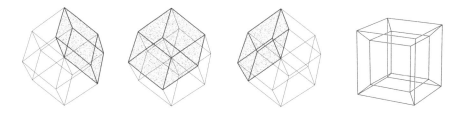

BOOK IV

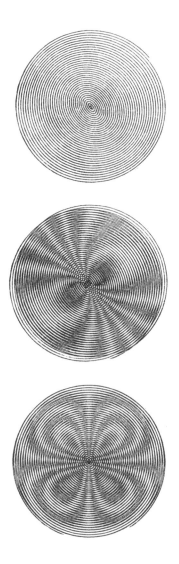

Unison (1:1): a spiral, a spiral drawn the same way over a
spiral, and a spiral drawn the opposite way over a spiral.

HARMONOGRAPH

A Visual Guide to the Mathematics of Music

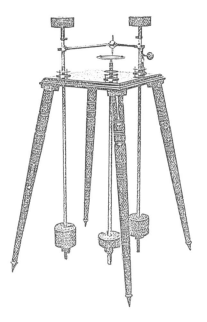

Anthony Ashton

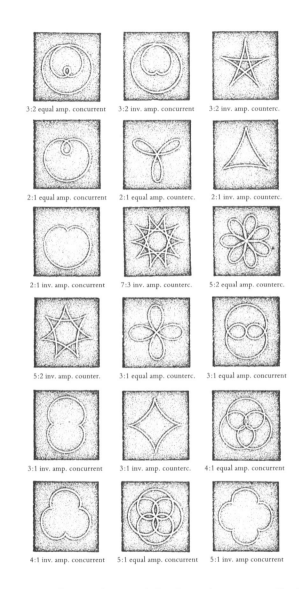

3:2 equal amp. concurrent	3:2 inv. amp. concurrent	3:2 inv. amp. counterc.
2:1 equal amp. concurrent	2:1 equal amp. counterc.	2:1 inv. amp. counterc.
2:1 inv. amp. concurrent	7:3 inv. amp. counterc.	5:2 equal amp. counterc.
5:2 inv. amp. counter.	3:1 equal amp. counterc.	3:1 equal amp. concurrent
3:1 inv. amp. concurrent	3:1 inv. amp. counterc.	4:1 equal amp. concurrent
4:1 inv. amp. concurrent	5:1 equal amp. concurrent	5:1 inv. amp concurrent

Harmonic patterns from Sir Thomas Bazley's Index to the Geometric Chuck *(1875), showing concurrent & countercurrent phases with equal & inverted amplitudes.*

INTRODUCTION

Many of the drawings in this book were produced by a simple scientific instrument known as a harmonograph, an invention attributed to a Professor Blackburn in 1844. Towards the end of the nineteenth century there seems to have been a vogue for these instruments. Victorian gentlemen and ladies would attend *soirées* or *conversaziones*, gathering round the instruments and exclaiming in wonder as they watched the beautiful and mysterious drawings appear. A shop in London sold portable models that could be folded into a case and taken to a party. There may well be some of these instruments hidden in lofts throughout the country.

From the moment I first saw drawings of this kind I was hooked: not only because of their strange beauty, but because they seemed to have a meaning—a meaning which became clearer and deeper as I found out how to make and operate a harmonograph. The instrument draws pictures of musical harmonies, linking sight and sound.

However, before going any further I feel I should issue a health warning. If you too are tempted to follow this path, beware! It is both fascinating and time-consuming.

I should acknowledge my debt to a book called *Harmonic Vibrations*. It was coming across this book in a library soon after the end of the second world war that introduced me to the harmonograph. Seeing that the book had been published by a firm of scientific instrument makers in Wigmore Street I went one day to see if they were still there. They were, though reduced merely to making and selling projectors.

I went into the shop and held up my library copy of the book for the elderly man behind the counter to see.

"Have you any copies of this book left?" I asked him.

He stared at me as though I was some sort of ghost, and shuffled away without a word, returning in a few minutes with a dusty, unbound copy of the book.

"That's marvelous," I said, "how much do you want for it?"

"Take it", he said, "it's our last copy, and we're closing down tomorrow."

So I have always felt that someday I must write this book.

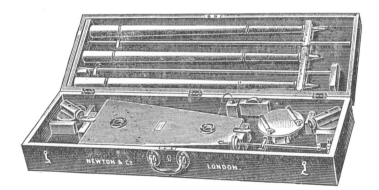

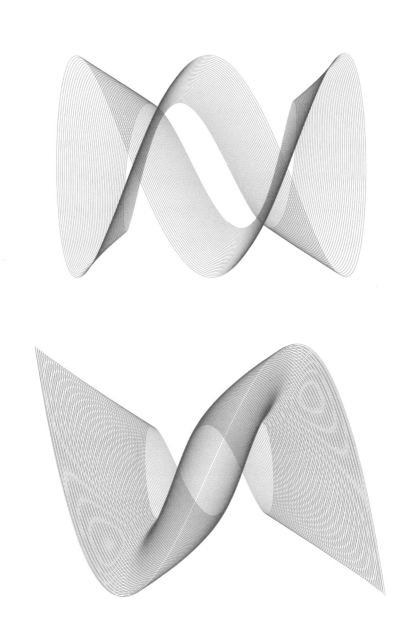

The Discovery of Harmony
on passing a blacksmith

To understand what the harmonograph does we need first to glance at the elements of musical theory.

Pythagoras, some 2,500 years ago, is credited with discovering that the pleasing experience of musical harmony comes when the ratio of the frequencies consists of simple numbers. A widely recounted story tells how taking a walk he passed a blacksmith's shop. Hearing familiar harmonies in the ringing tones of the hammers on the anvil, he went in and was able to determine that it was the weights of the hammers which were responsible for the relative notes.

A hammer weighing *half* as much as another sounded a note *twice* as high, an *octave* (2:1). A pair weighing 3:2 sounded beautiful, a *fifth* apart. Simple ratios made appealing sounds.

The picture opposite shows experiments the philosopher went on to make (from Gafurio's *Theorica Musice*, 1492), as he found that all simple musical instruments work in much the same way, whether they are struck, plucked or blown.

Deeply impressed by this link between music and number, Pythagoras drew the metaphysical conclusion that all nature consists of harmony arising from number, precursor to the modern physicist's assumption that nature conforms to laws expressed in mathematical form. Looking at the picture you will see that in every example, hammers, bells, cups, weights or pipes, the same numbers appear: 16, 12, 9, 8, 6, and 4. These numbers can be paired in quite a few ways, all of them pleasant to the ear, and, as we shall see, also pleasant to the eye.

THE MONOCHORD OF CREATION
a singular string theory

There are seven octaves in the keyboard of a piano and nearly eleven in the total range of sound heard by the average person. The highest note of each octave has a frequency twice that of the first so the frequencies increase *exponentially*, on a *scale* beginning at 16 cycles per second (16 *Hertz*) with the lowest organ note and ending with about 20,000 per second. Below 16 Hz we experience *rhythm*. A range of ten octaves represents about a thousandfold increase in frequency ($2^{10} \approx 10^3$).

There is a hint here of what we can think of as the 'great monochord' of the universe, also on a scale, this time stretching from a single quantum fluctuation at the bottom, to the observable universe at the top, passing through the various 'octaves' of atom, molecule, quantities of solid, liquid and gaseous matter, creatures great and small, planets, stars and galaxies. Here too the scale is exponential, but usually measured in powers of ten, and covering a range of more than 10^{40}.

Robert Fludd's 17th century engraving (*opposite*) tells a similar story: the musical scale follows the same exponential principle underlying the design of the universe.

						The Observable Universe
Proton	Cell	Mouse	Planet	Solar system		
Molecule	Insect	Whale	Star			Event horizon
Quarks		Forest			Galaxies	
Atom			Giant star	Star clusters	Galaxy clusters	
?	Nucleus	Grain of sand	Asteroid			?

Metres 10^{-20} 10^{-15} 10^{-10} 10^{-5} 10^{0} 10^{5} 10^{10} 10^{15} 10^{20} 10^{25}

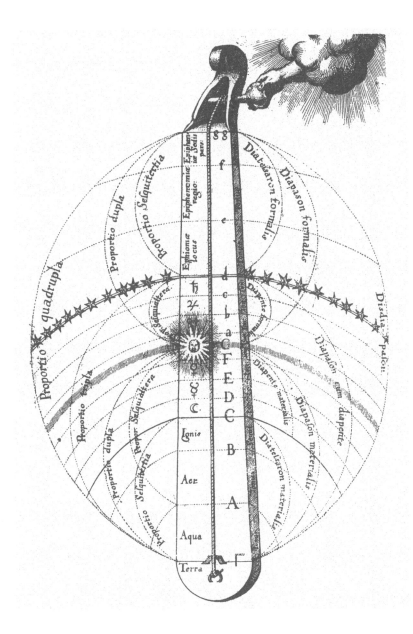

OVERTONES AND INTERVALS
harmonic ratios in and outside the octave

How are musical scales constructed? Listen very carefully as you pluck a string, and you will hear not only the main note, or *tonic*, but also a multitude of other harmonics, the *overtones*.

The principle is one of harmonic resonance, and affects not only strings and ringing hammers, but columns of air and plates too. Touching a string with a feather at the halfway or third point, as shown below, encourages regularly spaced stationary points, called *nodes*, and an overtone can be produced by bowing the shorter side. The first three overtones are shown opposite.

Musicians, however, need notes with intervals a little closer together than the overtone series, which harmonize *within* an octave. The lower diagram opposite shows the overtone series on the left, and the intervals developing within the octave on the right, in order of increasing dissonance, or complexity.

"All discord harmony not understood" wrote Alexander Pope. The brain seems to grasp easily the relationships implicit in simple harmonies, an achievement bringing pleasure; but with increasing complexity it falters and then fails, and failure is always unpleasant. For most people enjoyment fades as discord increases, towards the end of the series opposite. And, as we shall see, that is where the harmonograph drawings fade too.

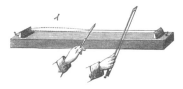
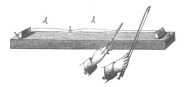

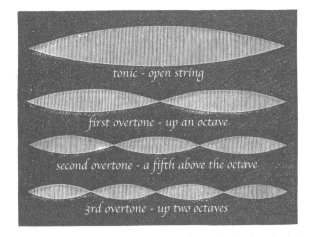

tonic - open string

first overtone - up an octave

second overtone - a fifth above the octave

3rd overtone - up two octaves

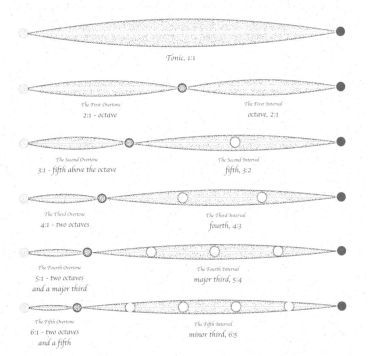

Tonic, 1:1

The First Overtone
2:1 - octave

The First Interval
octave, 2:1

The Second Overtone
3:1 - fifth above the octave

The Second Interval
fifth, 3:2

The Third Overtone
4:1 - two octaves

The Third Interval
fourth, 4:3

The Fourth Overtone
5:1 - two octaves
and a major third

The Fourth Interval
major third, 5:4

The Fifth Overtone
6:1 - two octaves
and a fifth

The Fifth Interval
minor third, 6:5

TONES AND HALFTONES
the fourth and fifth get their names

Pythagoras' hammers hide a set of relationships dominated by octaves (2:1), fifths (3:2) and fourths (4:3). The fifth and fourth combine to make an octave (3:2 × 4:3 = 2:1), and the difference between them (3:2 ÷ 4:3) is called a *tone*, value 9:8.

A natural pattern quickly evolves, producing seven discrete *nodes* (or *notes*) from the starting tone (or *tonic*), separated by two *halftones* and five tones, like the Sun, Moon and five planets of the ancient world.

The interval of the fifth (3:2), the leap to the *dominant*, naturally divides into a major third and minor third (3:2 = 5:4 × 6:5), the major third essentially consisting of two tones, and the minor third of a tone and a halftone. The thirds can be placed major before minor (*to give the major scale shown in the third row opposite*), or in other ways.

Depending on your harmonic moves, or *melody*, different *tunings* appear, for example two perfect tones (9:8 × 9:8 = 81:64) are *not* in fact the perfect major third 5:4, but are slightly sharp as 81:80 (the *syntonic* or *synoptic comma*, the Indian *shruti*, or *comma of Didymus*), more of which later.

Simple ratios, the octave and fifth, have given rise to a basic *scale*, a pattern of tones and halftones and, depending on where in the sequence you call home, seven *modes* are possible (*see page 382*).

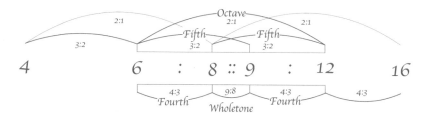

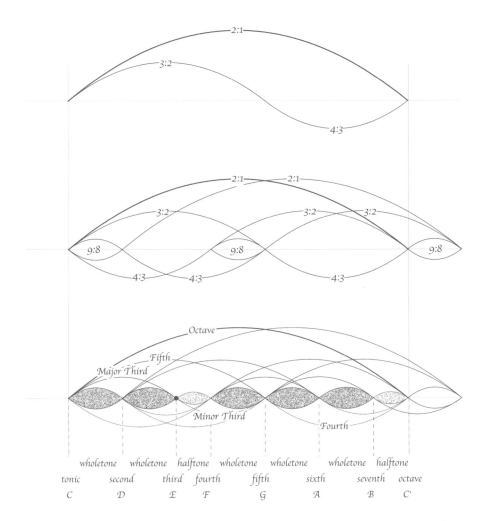

wholetone · wholetone · halftone · wholetone · wholetone · wholetone · halftone

tonic · second · third · fourth · fifth · sixth · seventh · octave

C · D · E · F · G · A · B · C'

Above: The basic manifestation of the major scale. In Pythagorean tuning all tones are exactly 9:8, creating the leimma halftone of 256:243 between its major third (81:64) and the perfect fourth (4:3). The sixth and the seventh are defined as successive perfect tones above the fifth.

In Diatonic tuning the major third is perfect at 5:4, which squeezes the second tone to 10:9 (a minor tone), leaving 16:15 as the diatonic halftone up to the fourth. The diatonic sixth is 5:3, a major third above the fourth, a minor tone above the fifth. The diatonic seventh (15:8) is a major tone above that, a major third above the fifth and a halftone below the octave.

DERIVING THE SCALE
monochord, tetrachord, and major chord

The pure ratios of fourth, fifth and octave make up the paradoxical trinity of the Pythagorean Tetrachord, simple yet curiously mystifying, in that here the two notes of the sounded octave become as one. Thus a fifth up is a fourth down, and vice versa (*see illustration, page 194*), and four notes, within the unity of the octave, become three. The reflexive relationship of these pure tones forms the basis of Western tonality, and from them is derived the major scale (*opposite top*).

When the fourth is dropped an octave, the fundamental, or tonic, now becomes its fifth. This results in the circle of fifths, in which each new tone is the second overtone (i.e., fifth) of its predecessor (*lower, opposite*). The fundamental, C, is flanked by the subdominant F, and dominant, G. When tonic and fourth are sounded together, the tonic, I, is undermined by, and almost subsumed into the subdominant, IV. The three constituent elements of the Tetrachord, in their delicate balance, have defined the way we hear and understand music.

Scales can be derived from various basic principles. The harmonic scale consists of the third octave of the overtone series, C D E F♯ G A B♭ B C. In Indian music Saraswati, Goddess of music and science, has a *raga* containing seven of these eight notes. Indian tunings, however, extend to 22 tones, or *shrutis*, to allow for the syntonic comma, 81:80 (*see page 194*), From these tones, seven notes are chosen. The Persian gamut of 17 tones includes the seven white notes on a piano plus the black notes, which are split down the middle, for sharps are *not* flats (*below*).

$$G^b \ D^b \ A^b \ E^b \ B^b \ F \ C \ G \ D \ A \ E \ B \ F^\# \ C^\# \ G^\# \ D^\# \ A^\#$$

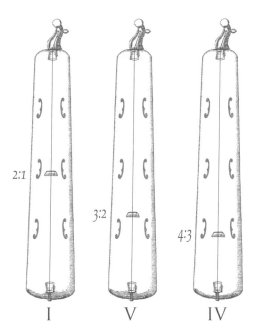

2:1

3:2

4:3

I   V   IV

Left: Using a monochord with a movable bridge, Pythagoras demonstrated the relationship between string length and musical tones. The most simple tones are the tonic, fifth, and fourth, also known as the Tetrachord. The early overtones of each of these yield the three major chords or triads, I, IV and V (e.g. the early overtones of C are C, G, C″, E″). The octaves merge acoustically. Thus, with C as root, or fundamental:

Chord I: C, E, G
Chord IV: F, A, C
Chord V: G, B, D

Combining these notes within the span of a single octave gives the major scale:
C, D, E, F, G, A, B, C

Right: Deriving the major scale from a circle of fifths requires starting from the subdominant F. With a circle of fifths beginning at C, bringing the first seven notes down into a single octave leads to a pure Lydian mode: C D E F♯ G A B C, known to the ancient Greeks as Syntolydian, and containing the 'raised' 4th, F♯. Interestingly, both the harmonic scale (opposite) and Lydian mode contain the difficult-sounding 'augmented fourth,' or tritone, known to medieval musicologists as the 'diabolus in musica'.

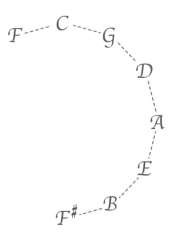

197

ARRANGING THE HARMONIES
the power of silence

The simple ratios of the primary overtones and undertones can be plotted on an ancient grid known as a *lambdoma* (*opposite top*), after the greek letter λ. Some intervals have the same value (e.g., 8:4 = 6:3 = 4:2 = 2:1), and if lines are drawn through these it quickly becomes apparent that the identities converge on the silent and mysterious ratio 0:0, which is 'outside the diagram'.

A further contemplative device used by the Pythagoreans was the *Tetraktys*, a triangular arrangement of ten elements in four rows (1 + 2 + 3 + 4 = 10). The basic form is given opposite lower left, the first three rows producing the simple intervals. In another lambdoma (*lower, opposite right*), numbers are doubled down the left side and tripled down the right, creating tones horizontally separated from their neighbours by perfect fifths. After the trinity (1, 2 and 3) notice the numbers produced, 4, 6, 8, 9, 12, and then look again at the picture on page 189.

Below we see a fuller range of monochord positions, with the overtones on the right, and intervals on the left (*see pages 193 and 195*).

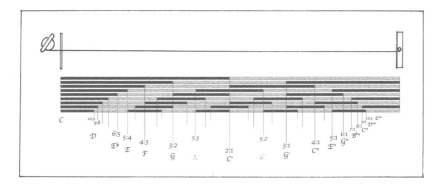

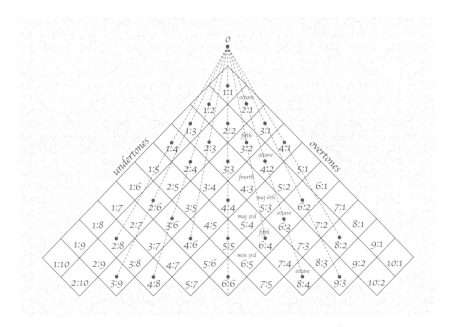

Pythagorean and medieval tunings, called 3-limit, recognized no true intervals except for ratios involving 1, 2 and 3. The lambdoma below right expresses this numerically as any element relates to any neighbour by ratios only involving 1, 2 and 3, so we can move around by octaves and fifths. Squares ($4=2^2$, $9=3^2$) and cubic volumes ($8=2^3$, $27=3^3$) also appear. Add further rows and the numbers for the Pythagorean scale soon appear, 1 9:8 64:81 4:3 3:2 27:16 16:9 2:1. This has four fifths and five fourths but no perfect thirds or sixths. These came later with the diatonic scale and its perfect thirds (6:5:4) as polyphony and chords slowly took over from plainchant and drone.

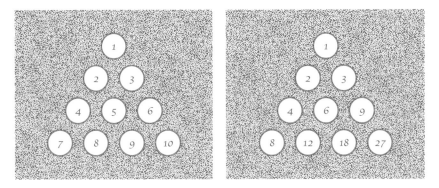

Lissajous Figures

sound made shape

In the mid nineteenth century a French mathematician, Jules Lissajous, devised an experiment: He found that if a small mirror was placed at the tip of a tuning fork, and a light beam aimed at it, then the vibration could be thrown on to a dark screen. When the tuning fork was struck, a small vertical line was produced and if quickly cast sideways with another mirror it produced a sine-wave (*below*).

Lissajous wondered what would happen if instead of casting the wave sideways he were to place another tuning fork at right angles to the first to give the lateral motion. He found that tuning forks with relative frequencies in simple ratios produced beautiful shapes, now known as Lissajous figures.

On the screen (*opposite top*), we see the octave (2:1) as a figure of eight, and below it various *phases* of the major and minor third. These were some of the first fleeting pictures of harmony, which were doubtless familiar to Professor Blackburn when he devised the harmonograph.

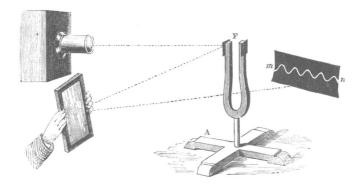

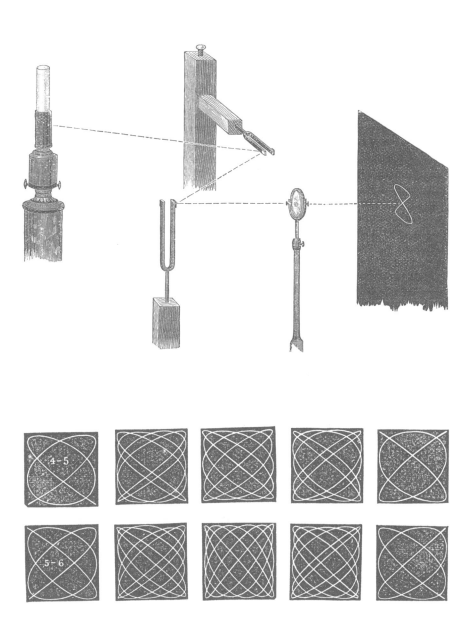

THE PENDULUM
keeping time

A fundamental law of physics (in one formulation) states that left to itself any closed system will always change towards a state of equilibrium from which no further change is possible.

A pendulum is a good example. Pulled off center to start, it is in a state of extreme disequilibrium. Released, the momentum of its swing carries it through nearly to the same point on the other side. As it swings it loses energy in the form of heat from friction at the fulcrum and brushing against the air. Eventually the pendulum runs down, finally coming to rest in a state of equilibrium at the center of its swing.

Going back 500 years, Galileo, watching a swinging lamp in the cathedral of Pisa, realized the frequency of a pendulum's beat depends on its length: the longer the pendulum the lower the frequency. So the frequency can be varied at will by fixing the weight at different heights. Most importantly, as the pendulum runs down, the frequency stays the same.

Here, therefore, is a perfect way to represent a musical tone, slowed down by a factor of about a thousand to the level of human visual perception. For a simple harmonograph two pendulums are used to represent a harmony, one with the weight kept at its lowest point, while the weight on the other is moved to wherever it will produce the required ratio.

As we shall see, the harmonograph combines these two vibrations into a single drawing, just as two musical tones sounded together produce a single complex sound.

The theoretical length of the variable pendulum that will produce each harmony can be calculated, for the frequency of a pendulum varies inversely with the square root of its length. This means that while the frequency doubles within the octave, the length of the pendulum is reduced by a factor of four.

Figures are given for a pendulum 80cm long, a convenient length for a harmonograph. These theoretical markers provide useful 'sighting shots' for most of the harmonies. Note that the pendulum length is measured from the fulcrum to the center of the weight.

Interval Name	Approx. Note	Diatonic Ratio	Length (cm)	Freq. (s¹)
Octave	C'	2:1	20	66.0
Maj. 7th	B	15:8	22.8	62.8
Min. 7th	B♭	9:5	24.7	59.4
Maj. 6th	A	5:3	28.8	55.8
Min. 6th	G♯	8:5	31.2	53.6
5th	G	3:2	35.6	50.3
4th	F	4:3	45.0	44.7
Maj. 3rd	E	5:4	51.2	41.9
Min. 3rd	E♭	6:5	55.6	40.2
2nd	D	9:8	63.2	37.7
Halftone	C♯	16:15	70.3	35.7
Unison	C	1:1	80	33

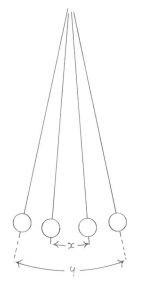

When a pendulum is pulled back and then released, the weight tries to fall towards the center of the Earth, accelerating as it does so. As the pendulum runs down, the rate of acceleration, and so the speed of travel, is reduced, but in equal proportion to the distance of travel. The result is that the period (the time taken for two beats) or the number of periods in a given unit of time (the frequency) remains unchanged. In the picture to the left the frequencies of beats x and y are the same. For the pendulum formula, see page 383.

TWO HARMONOGRAPHS
lateral and rotary

_____

In the simplest version of the instrument two pendulums are suspended through holes in a table, swinging at right angles to each other. Projected above the table, the shaft of one pendulum carries a small platform with a piece of paper clipped to it, while the shaft of the other pendulum carries an arm with a pen.

As the pendulums swing the pen makes a drawing that is the result of their combined motion (*see left side opposite*). Both pendulums begin with the same length, further drawings can be obtained as one is then shortened by sliding the weight upwards and securing it with a clamp at various points. The harmonic ratios can be displayed in turn.

By using three pendulums however, two circular, or *rotary*, movements can be combined, with fascinating results (*see right side opposite*). Two of the pendulums swing at right angles as before, but are now both connected by arms to the pen, which in all rotary designs describes a simple circle.

Situated under the circling pen, the third and variable pendulum is mounted on gimbals, a device familiar to anyone who has had to use a compass or cooking stove at sea. Here it acts as a rotary bearing, enabling the pendulum carrying the table to swing in a second circle under the pen. As the pen is lowered the two circles are combined on the paper.

A further source of variation is also introduced here, for the two circular motions can swing in the same (concurrent) or opposite (countercurrent) directions, producing astonishing drawings with very different characteristics (*see appendices on pages 384–385*).

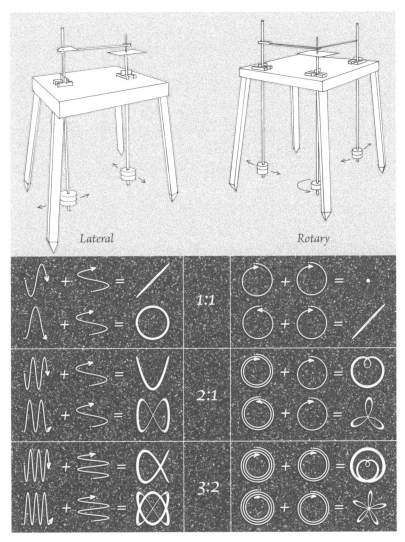

Lateral

Rotary

1:1

2:1

3:2

Above: Two harmonographs and some of the simple patterns they draw. On the left the simple
lateral version and its patterns (open and closed phase); on the right the three-pendulum, rotary
harmonograph and its drawings (concurrent and countercurrent). See too the twin-elliptic
harmonograph illustrated in the lower right-hand corner of page 387.

SIMPLE UNISON 1:1
and the arrow of time

The simplest harmonograph drawing is produced when both pendulums are the same length and the table is stationary. With the pen held off the paper both pendulums are pulled back to their highest points. One is released, followed by the other when the first is at its mid-point. The pen is then lowered on to the paper to produce a circle developing into a single spiral.

If the two pendulums are released together then the result will be a straight diagonal line across the paper, the 'closed' *phase* of the harmony, as opposed to the circular 'open' phase. At intermediate phase points elliptical forms appear (*below*).

The running-down of harmonograph pendulums is an exact parallel to the fading of musical notes produced by plucked strings, and can also be thought of as graphically representing the 'arrow of time' (*see opposite*), with the unchanging ratios of the frequencies representing the eternal character of natural law. The characteristics of the drawings result from the meeting of the running-down process with the 'laws' represented by the various frequency ratios. We see that music, like the world, is formed from unchanging mathematical principles deployed in time, creating complexity, variety and beauty.

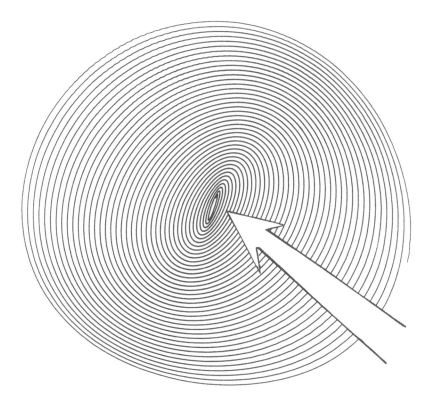

The inexorable direction of change, linked to the asymmetry of time (before-now-after), was vividly described by the scientist Arthur Eddington (1882–1944) as 'the arrow of time'. Throughout the process of continuing universal degradation, the dwindling stock of 'useful' energy encounters a hierarchy of fixed physical laws conforming to mathematical formulae, and it is from the interaction of these unchanging laws with the arrow of time that comes a changing world of astonishing complexity, variety and beauty. The pendulum runs down from a state of disequilibrium to one of equilibrium, and the same is true, we are told, of the universe, the ultimate 'closed system'. From a state of extreme disequilibrium it plunged via the 'Big Bang' towards its future ultimate state of utterly dark, frozen equilibrium. Between the beginning and the end there is a continual, cumulative transformation of 'useful' energy, capable of forming temporary structures and causing events, into 'useless' energy forever lost.

Near Unison
lateral phases and beat frequencies

A source of pleasing variety in harmonograph drawings comes from small departures from perfect harmonies. This seems to involve a principle widespread in nature as well as in the work of many artists. There is a particular charm in the near-miss.

An example from music suggests itself here. When two notes are sounded in near unison, the slight difference in their frequencies can often add richness or character to the sound. The two reeds producing a single note in a piano accordion have slightly different frequencies, the small departure from unison causing 'beats', a subtle warbling or throbbing sound (*see page 383*).

Set the weights for unison and then shorten the variable pendulum slightly. Swing the pendulums in open phase, producing a circle turning into an increasingly narrow ellipse and then a line. If the pen is allowed to continue, the line will change into a widening ellipse, a circle, and a line again at right angles to the first. And so on. The instrument is working its way through the phases of unison shown on page 206.

If the variable pendulum is then further shortened in stages, a series of drawings like those opposite will be produced. The repetitive pattern represents 'beats' with increasing frequency as the discrepancy between the notes widens. Eventually the series fades into a scribble that is a fair representation of discord, though even here there is a hint of some higher-number pattern.

For most people this fading of visual harmony occurs at about the same point as the audible harmonies fade.

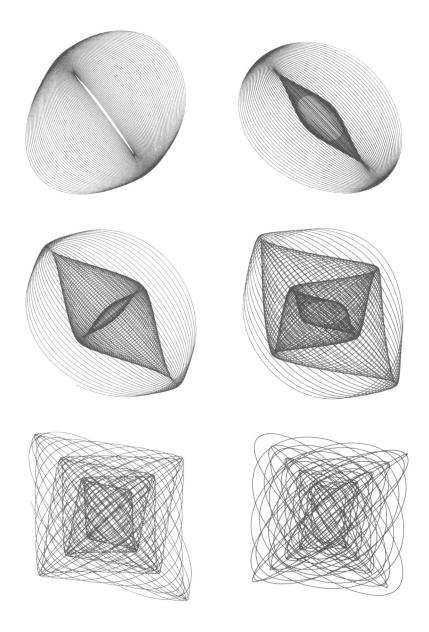

ROTARY UNISON 1:1
eggs and shells

At first this is disappointing: unison in contrary motion produces a straight line across the paper, like the closed phase of lateral unison. From concurrent motion there comes a mere dot that turns into a line struggling towards the center, pen and paper going round together.

However, changing to near-unison is richly rewarding. In contrary motion come a variety of beautiful, often shell-like, forms with fine cross hatchings. For best results lift the pen off the paper well before the pendulums reach equilibrium.

Surprisingly, from concurrent near-miss motion there come various spherical or egg-shaped forms. To produce an 'egg' the pen should be lowered when it is dawdling at the center. It then spirals its way outwards, reaching a limit before returning as the pendulums run down. Because the lines toward the perimeter get closer together, the drawing appears three-dimensional.

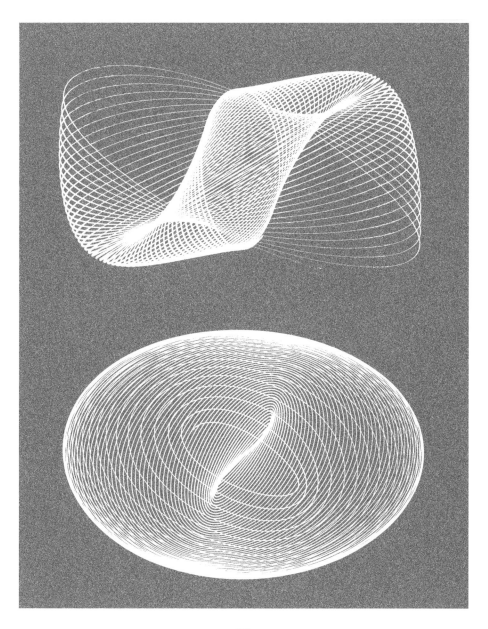

THE LATERAL OCTAVE 2:1

figures of eight and wings

After unison the next harmony to try is the octave. Here there is a technical difficulty, for the variable pendulum has to be very short, and because of the greater amount of friction involved runs down quickly. The trick is to add a weight to the top of the invariable pendulum, which slows it down (*see page 183*). The variable pendulum can then be longer.

Unfortunately this means that for the octave, and other ratios where one pendulum is going much faster than the other, the theoretical markers have to be ignored, and the right point found by trial and error.

With one pendulum beating twice as fast and at right angles to the other, the octave in open phase takes the form of a figure-of-eight (a coincidence), repeated in diminishing size as the pendulum runs down.

If both pendulums are released at the same time to produce the closed phase, the result is a cup-shaped line that develops into a beautiful winged form with fine cross-hatchings and interference patterns. Small adjustments produce striking variations.

The octave is the first overtone (*see page 192*).

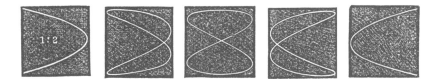

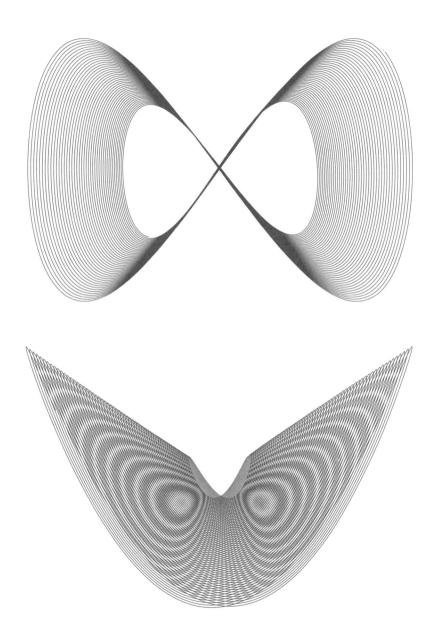

THE ROTARY OCTAVE 2:1

hearts and triangles

From rotary motion with a 2:1 ratio come some of the most beautiful of all harmonograph drawings: simple, graceful and often surprising. Remember, all that is happening here is that two circular motions, one almost exactly twice as fast as the other, are being added together.

Contrary motion produces a trefoil shape with many fine variations (*right hand images opposite*). Starting with a smaller size or *amplitude* in the faster rotation produces a triangle, or pyramid.

The octave in concurrent motion produces a heart-shaped form with a simple inner loop (*below left and left hand column opposite*). Here there is a link with the ancient tradition of the music of the spheres, for this is the shape an observer on Uranus would ascribe to the movement of Neptune, or vice-versa. This is because the planets orbit the Sun concurrently, Uranus in 84 years and Neptune in 165, approximately performing an octave. The planet Mercury sings a perfect octave all by itself, as one of its days is two of its years (*see Book VI in this volume for much more on relationships such as these*).

Near-misses in the ratios of rotary drawings set the designs spinning (*lower, opposite row*).

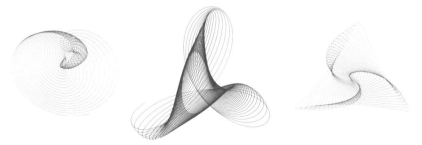

THE LATERAL FIFTH 3:2
and the second overtone 3:1

Next to be tried is the harmony of the fifth, intermediate between the simplicity of unison and octave and the more complex harmonies that follow.

It will be seen from the open phase drawing opposite that the fifth has three loops along the horizontal side and two along the vertical. The number of loops on each side gives the ratio, 3:2. Looking back at the octave, there are two loops to one, and with unison there is only one 'loop', however you look at it. This is the general rule for all lateral harmonograph ratios, and if a harmony appears unexpectedly during experiments, it can usually be identified by counting the loops on two adjacent sides.

The fifth also appears as 3:1, the second overtone, a fifth above the octave (*see open and closed phase drawings of 3:1 on page 187*). Drawing ratios outside the octave may require a twin-elliptic harmonograph (*see page 387*). The phase-shifted pair below are stereographic; if you go cross-eyed they will jump into 3-D.

 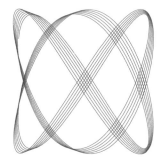

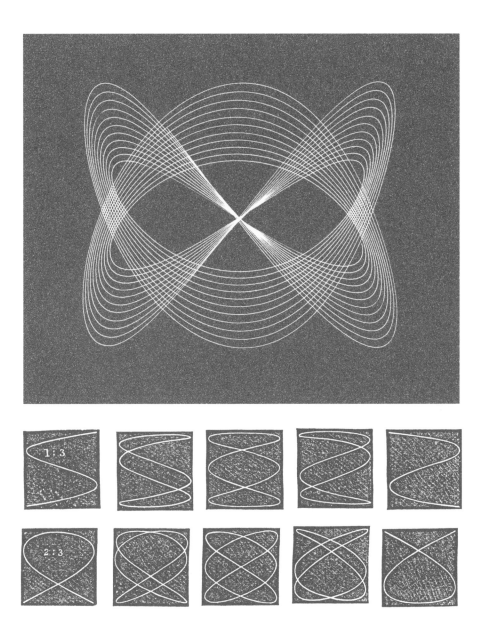

THE ROTARY FIFTH 3:2
encircled hearts and fives

The 'loudness' of musical tones is represented on the harmonograph by *amplitude*, the relative sizes of the two circular motions. In rotary drawings this is much more important than phase, which simply orients the whole design on the page.

The third drawing below shows a rotary fifth in contrary motion where the higher-frequency, faster moving, pendulum has much the wider swing. In the 'spiky' drawing to its right it is the other way round. At equal amplitude all lines pass through the center (*see table on page 385*).

The top four drawings opposite show rotary forms of 3:2, concurrent on the left, and countercurrent on the right. The second row shows the effect of a near-miss in the harmony, which makes the patterns spin.

The lower two images opposite are from the 1908 book *Harmonic Vibrations*. They show the second overtone, 3:1, a fifth above the octave (3:1 = 2:1 × 3:2), concurrent on the left, countercurrent on the right.

With concurrent pictures, the number of swirls in the middle is given by the difference between the two numbers of the ratio. So the concurrent patterns for the primary musical intervals 2:1, 3:2, 4:3, 5:4, and 5:6 all have a single heart at their center.

THE FOURTH 4:3

with thirds, sixths, and sevenths

By now it will be evident that each harmony displays its own distinct aesthetic character. Unison is simple and assertive. The octave introduces an emphatic flourish, and the fifth, while still fairly simple, has added elegance.

With the fourth the pattern becomes more complicated, though the design is still recognisable without counting the loops. The upper diagram opposite shows the fourth in open phase, the lower in closed phase. An increasing sophistication becomes apparent, and some of the closed phase and near-miss variants have a strange exotic quality.

Introducing the perfect thirds of diatonic tuning increases the complexity. The major third (5:4) is found below the fourth, the interval between them, a *diatonic halftone*, working out as 4:3 ÷ 5:4 = 16:15. A fourth and a major third (4:3 × 5:4) produce the *major sixth*, 5:3, a minor third (6:5) below the octave and a *minor tone* (10:9) above the fifth. Likewise, a fourth and a minor third (4:3 × 6:5) create the *minor sixth* (8:5), a major third (5:4) below the octave and a halftone (16:15) above the fifth.

A fifth and a major third (3:2 × 5:4) produce the *major seventh*, 15:8, while a fifth and a minor third (3:2 × 6:5) give the *minor seventh*, 9:5. These are the elements of the diatonic, or *just*, scale.

FURTHER HARMONICS
seven limit and higher number ratios

As the numbers in the ratios increase it becomes harder to distinguish the harmonies one from another at a glance: the loops have to be counted, and slight variations produce little of aesthetic value. A typical example, 7:5, is shown opposite top.

Rotary motion produces a series of increasingly complex drawings, influenced by relative frequency, amplitude and direction. In contrary motion the total number of loops equals the sum of the two numbers of the ratio. With concurrent motion the nodes turn inwards, and their number is equal to the difference between the two numbers of the ratio.

The contrary drawings below show a fourth (4:3), another fourth, a major sixth (5:3) and a major third (5:4). The lower pictures opposite, drawn over a hundred years ago, show unequal amplitude drawings of the perfect eleventh 8:3 (an octave and a fourth) and the ratio 7:3 which is found in seven-limit jazz tuning (not covered in this book).

Two octaves and a major third (4:1 × 5:4) equal 5:1, the fourth overtone, which differs from four fifths $(3:2)^4$ as our friend 80:81, the syntonic comma (*see page 194*). In *mean tone* tuning, popular during the Renaissance, this misfit was ironed out and the fifths were flattened very slightly, to $5^{1/4}$ or 1.4953, falling out of tune to please the thirds and sixths.

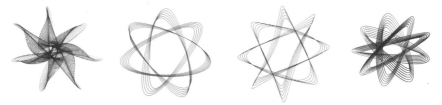

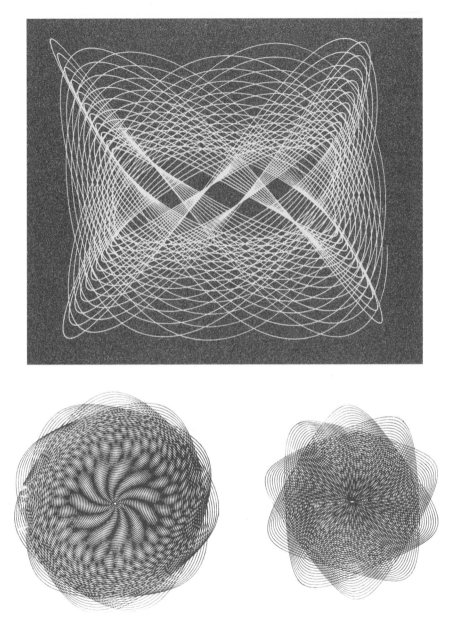

Amplitude

circles, polygons, flowers and another circle

Much variation can be obtained from a rotary ratio by having unequal sizes in the two circular motions. Opposite we see two frequencies related by a major sixth (5:3). A lower frequency note begins to be influenced by, combines with, and is then more or less replaced by a higher frequency one. When the two notes are at equal volume the lines all pass through the center (*see pages 386–387*). Notice that the sequence is not symmetrical.

Below we see the first three overtones. For the spikiest shapes simply invert the amplitudes. For polygons, square them first.

If you have ever played with a 'Spirograph', the harmony is determined by the cogging ratio, and it is the amplitude which is adjusted when you change penholes on the wheel.

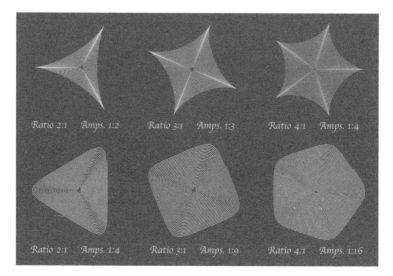

Ratio 2:1   Amps. 1:2   Ratio 3:1   Amps. 1:3   Ratio 4:1   Amps. 1:4

Ratio 2:1   Amps. 1:4   Ratio 3:1   Amps. 1:9   Ratio 4:1   Amps. 1:16

TUNING TROUBLES
the Pythagorean comma

Leaving the harmonograph drawings and returning to the principles of music, you may have noted that musical intervals do not always agree with one another. A famous example of this is the relationship between the octave and the perfect fifth (3:2).

In the central picture opposite, a note is sounded in the middle at o, and moved up by perfect fifths to give the sequence c, g, d, a, e, etc. (*numbered opposite, each turn of the spiral representing a perfect octave*). After twelve fifths we have gone up seven octaves, but the picture shows that we have overshot the final octave slightly, and gone sharp. This is because $(\frac{3}{2})^{12} \approx 129.75$, whereas $(2)^7 = 128$. The difference is known as the Pythagorean comma, proportionally 1.013643, approximately 74:73.

If you kept on spiralling you would eventually discover, as the Chinese did long ago, that 53 perfect fifths (or *Lü*) almost exactly equal 31 octaves. The first five fifths produce the pattern of the black notes on a piano, the Eastern *pentatonic* scale (*see pages 272 and 380*).

The smaller pictures opposite show repeated progressions of the major third (5:4), the minor third (6:5), the fourth (4:3), and the whole tone (9:8), all compared to an invariant octave.

It's strange. With all this harmonious interplay of numbers you would have expected the whole system to be a precisely coherent whole. It isn't. There are echoes here from the scientific view of a world formed by broken symmetry, subject to quantum uncertainty and (so far) defying a precise comprehensive 'theory of everything'. Is this why the 'near miss' is so often more beautiful than perfection?

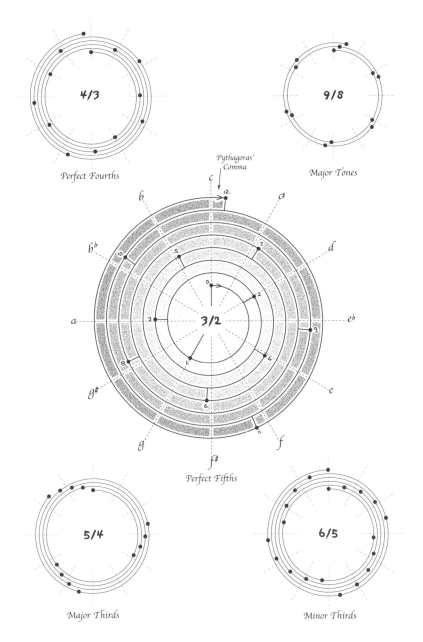

4/3

Perfect Fourths

9/8

Major Tones

Pythagoras'
Comma

3/2

Perfect Fifths

5/4

Major Thirds

6/5

Minor Thirds

EQUAL TEMPERAMENT
changing keys made easy

Although early tunings enabled many pure harmonics to be played, it was hard to move into other keys, one could only really change *mode* (*see page 382*). Musicians often had to retune their instruments, or use extra notes reserved for specific scales (classical Indian tuning uses 22 notes).

In the sixteenth century a new tuning was developed which revolutionized Western music and which predominates today. The octave is divided into twelve fixed *equal* intervals, each *chromatic halftone* being 1.05946 times its neighbour ($2^{1/12}$, roughly 18:17).

The twelve equally spaced notes are arranged in a circle below. Six (flat) wholetones now make an octave, as do four (very flat) minor thirds, or three (sharp) major thirds. The Pythagorean comma vanishes, as do *all* perfect intervals except the octave—it's a clever fudge which allows us to change key easily. It is slightly 'out of tune' and we hear it every day.

Triads are chords of three notes. Opposite top we see major and minor triads involving the note c, in the key of c. Use the mastergrid (*opposite, below*) to navigate the equal-tempered sea, and perceive any 3-4-5 triad (a chord of three notes) in three distinct keys (*after Malcolm Stewart*).

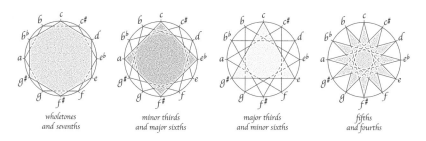

wholetones
and sevenths

minor thirds
and major sixths

major thirds
and minor sixths

fifths
and fourths

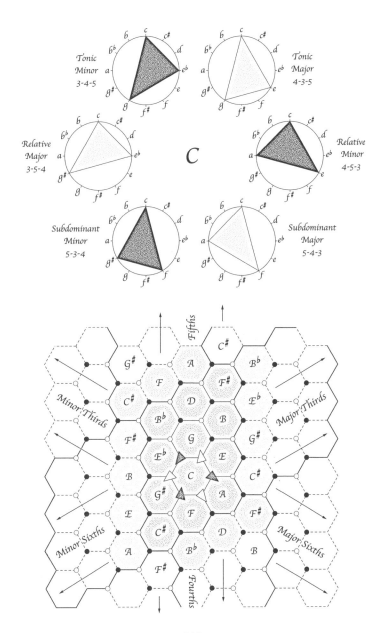

THE KALEIDOPHONE
squiggles from a vibrating rod

Despite the invention of equal temperament, scientists continued to investigate pure ratio harmonics. An interesting nineteenth century precursor to the harmonograph was the *kaleidophone*, invented by Sir Charles Wheatstone in 1827. Like the harmonograph, it displayed images of harmonics.

The simplest version of the device consists of a steel rod with one end firmly fixed into a heavy brass stand and the other fixed to a small silvered glass bead, so that when illuminated by a spotlight a bright spot of light is thrown upon a screen placed in front of it. Depending on how the kaleidophone is first struck, and then subsequently stroked with a violin bow, a surprising number of patterns can be produced (*a few are shown opposite*).

The kaleidophone does not behave like a string, as it is only fixed at one end. Like wind instruments, which are normally open at one end, the mathematics of its harmonics and overtones are slightly more complicated than the monochord or the harmonograph and the positions of the nodes are more variable (*the lower images opposite show some early overtones*).

Other versions of the kaleidophone used steel rods with square or oval cross-sections to give further patterns. Wheatstone used to refer to his invention as a 'philosophical toy', and indeed, as we look at these patterns, it is easy to feel wonder at their simple beauty.

To make your own kaleidophone, try fixing a knitting needle into a vice and sticking a silver bead or cake decoration ball to the free end. Use or make a bright point light source.

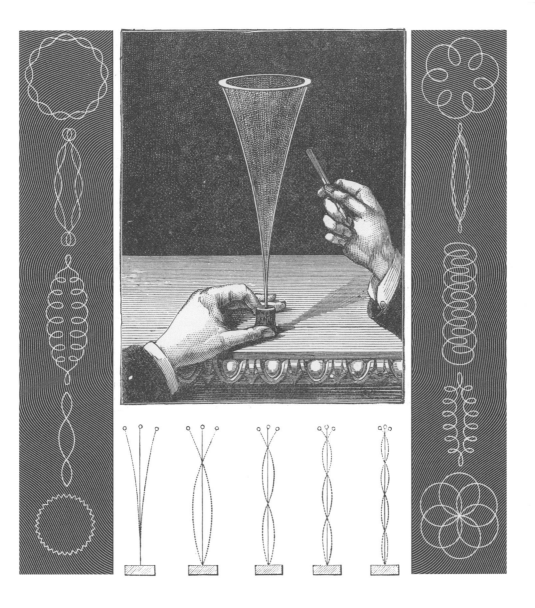

CHLADNI PATTERNS
vibrating surfaces

So far we have only considered vibrating strings and other simple systems, but surfaces also can be made to vibrate, and they too can display harmonic or resonant patterns.

In 1787 Ernst Chladni found that if he scattered fine sand on to a square plate, and bowed or otherwise vibrated it, then certain notes, generally harmonics of each other, each gave rise to different patterns in the sand on the plate. As with the harmonograph, other disharmonic tones produced a chaotic mess. Sometimes he found that further patterns could be created by touching the side of the plate at harmonic divisions of its length (*shown below*). This created a stationary node (*like the feather on page 192*). Later work revealed that circular plates gave circular patterns, triangular plates triangular patterns and so on.

The six pictures opposite are from Hans Jenny's book *Cymatics*, one of the seminal texts on this subject. The vibration picture appears gradually, the sand finding its way to stationary parts of the plate as the volume steadily increases through the sequence.

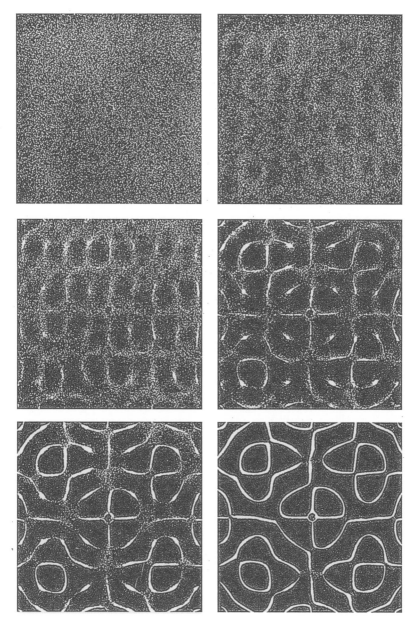

233

RESONANCE PICTURES
and how to sing a daisy

A more complete set of Chladni figures is shown opposite, all two or fourfold because they were produced on a square plate.

Below, however, we see some circular pictures. They were photographed by Margaret Watts Hughes, a keen singer, in the 1880s on an ingenious device called an *eidophone*, which consisted of a hollow base with a membrane stretched across it and a tube attached to its base with a mouthpiece at the other end. As Mrs. Hughes sang diatonic scales down the tube, fine lycopodium powder scattered on the taut membrane suddenly came to life, bouncing away from some places and staying still at others, producing shapes which she likened to various flowers.

Yet again, we see recognisable forms and shapes appearing from simple resonance and harmony.

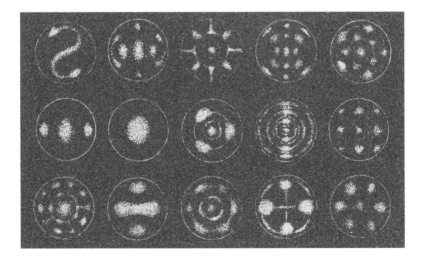

BOOK V

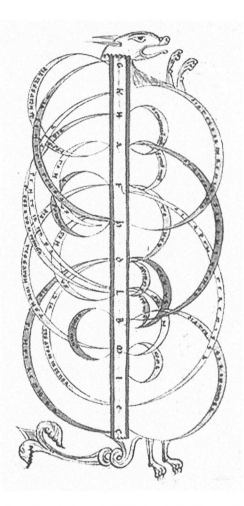

Zoomorphic musical diagram, from De Musica *by Boethius (480–524 AD).*

THE ELEMENTS OF
MUSIC

Melody, Rhythm, & Harmony

Jason Martineau

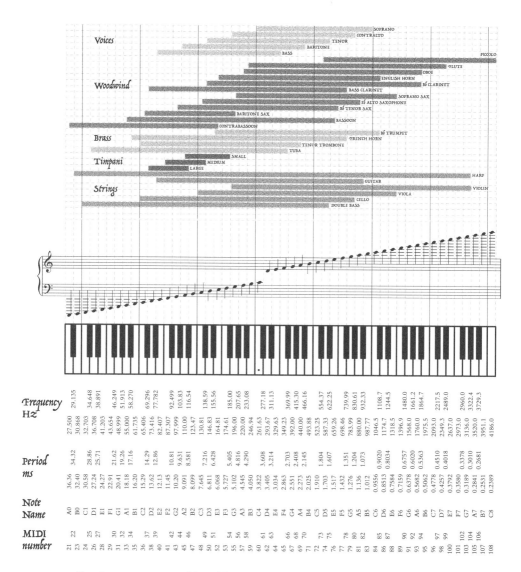

Above: Just over seven octaves of the audible spectrum, centered on middle C, showing the positioning of different instruments within the range of audible frequencies. Data for frequency and period is given for equal temperament.

INTRODUCTION

Music is the art medium that communicates interiority, being only perceived by the ears, and received by the mind. A strict approach to understanding music will consequently always have something lacking as music theory, in essence, is primarily descriptive and not prescriptive. The tendencies and practices in music are only observed and cataloged upon analysis, after the fact. It is the hearts and minds of human beings that shape and weave melodies, harmonies, and rhythms together into meaningful tapestries, imbued with the interior landscapes of their immediate experiences.

Much of the theory in this book is based upon the European classical tradition, starting around the early 18th century. The content is designed to get you started in understanding the relationships of the tones and rhythms, and in unpacking the inherent properties of sound in the process, and then, perhaps, music in general.

For the purposes of this book all principles are presented assuming equal temperament, the prevalent tuning system for over 300 years. The word 'tone' and 'note' may sometimes be used interchangeably, but generally 'tone' refers to the audible sound, and 'note' refers to the written symbol. Other terms can be consulted in the glossary.

I hope this book will reveal how the underlying harmonic template of sound acts as an organizational framework from which the fabric of music is woven, influencing our perception of accord, discord, tension and release, telling a story, making a journey.

WHAT IS MUSIC?

and all that jazz

Music is ... a mother's lullaby. It gives sound to our feelings when we
have no voice, words when we are silent. In it we praise, love, hope,
and remember. In the breath of the soul, the contours of the path of a
hummingbird in flight, and the wind that carries it; music shapes and
shivers into endless colours, nuanced and diverse, and eternally creative.
It is Spirit taking form.

Music is carried by the vibrations of molecules of air, like waves
upon an ocean. It perhaps uniquely captures and conveys the interior
landscape of one human mind to another, holding our tears and sweat,
pain and pleasure, packaged as paeans and preludes and etudes and
nocturnes. It is the texturization of the deliquescence of time, the ebb
and flow of mood and meaning. It ruminates, vacillates, contemplates,
and stimulates.

In music we organize and fantasize, arranging the elements of
music—*melody, rhythm,* and *harmony*—into meaningful shapes and
patterns. Its rhythms move our hands, feet and bodies to the pulses of
the universe. Its harmonies breathe with the exploratory intricacies
and curiosities of relationship and proportion, consonance, dissonance,
assonance, and resonance. Its melodies flitter into flights of fancy,
weaving woe and wonder.

When music is married to language, then what is spoken becomes
song, elevating the intentions and entreating us to listen more deeply,
making the profane sacred. Music soothes the soul, and the savage
beast. Orpheus mystifies creatures and trees, changing the course of
rivers, outplaying the Sirens' song with his lyre. Radha and Krishna
play the flute and dance jubilantly.

Temple Of Music by Robert Fludd (1574–1637). At lower left Pythagoras discovers that the ratios of hammer weights correspond to the octave, fifth, and fourth. On the wall lower right the basic rhythms are notated along a bass staff. Above these are the three kinds of hexachords, the six-note scales of medieval music. To the left of this is the lambda, a matrix of note relationships and ratios going back to Plato. The upper diagonal portion of that region is a diagram of intervallic distances between pitches in the scale. Running vertically further left is the monochord. Above the diagonal matrices is an actual composition putting all these practices into a musical structure.

EPIGRAMS AND DIALECTICS
ideas in sound

When music parallels language, it often chooses devices that resemble epigrams or poetic devices. Take the epigram 'Live, Love, Learn.' This collection of words, when arranged together, takes on an emergent inter-associative meaning that transcends the individual parts. Notice the alliteration of 'L's, and the use of 'learn' in the sequence to diffuse the rhyming scheme of the first two words, and close the set. Additionally, all three words are monosyllabic, and can be used both as conceptual infinitives (to live, to love, to learn) and as imperatives (Live! Love! Learn!).

In both music and language the components of epigrams are often synthesized or unified through paradox, an essential quality for having an aesthetic response and remembering the phrase. In music notes rise and fall, are consonant and dissonant, staccato and legato, or push and pull one another, these fundamental dualities representing the paradoxical nature of reality itself. Small pieces of meaning are arranged from them into larger forms based on their structure, and, perceiving this unity of opposites, the listener is temporarily removed from the dualistic separated stream of everyday life into the realm of unification.

The 'Happy Birthday' melody epigram (*see opposite*) undergoes various permutations and transformations but retains its fundamental characteristics, and thus stays satisfyingly recognizable to the end.

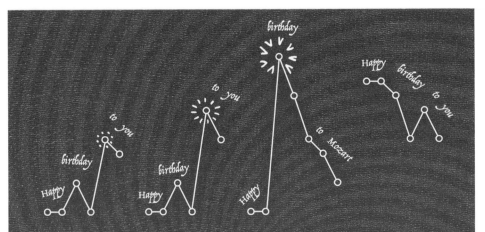

Above: 'Happy Birthday'. The melody starts on the 5th of the major scale, a long note followed by a short, repeated; then the 6th to the 5th, and a leap upwards of a fourth to the octave, before falling a half step. The first repetition of this idea is the same, except the leap upward is now larger (a fifth) followed by a fall of a whole step. The third time it is followed by a leap upwards of an entire octave (small interval becomes large), and where the melody previously fell by one step, now it falls twice, with a salient dissonance on the person's name elongated on a strong beat. The final phrase repeats the opening lyric, on a dissonant tone, falling by a step, leaping down, and another step. A well-constructed dialectic of leaps and steps, up and down, resting and moving tones, alternating on strong and weak beats, carrying us effortlessly on a tiny journey, a slow striving ascent upward, and a gentle cascade downward, parachuting to the tonic or home key of the piece.

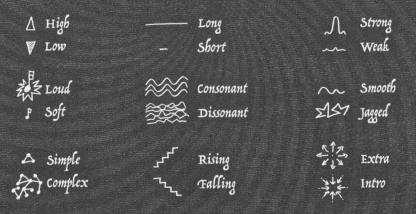

Above: Much of the communication of meaning in music is dependent upon opposites. Chromatic-diatonic, leap-step, repeat-contrast … opposites interplay as the narrative unfolds to carry the listener along. Contexts are defined immediately by the first sounds the music presents, and all that follows is constantly compared to what precedes, in the short-term and long-term, as time passes. Thus the interplay of these dialectics becomes clear as the music unfolds. Opposites in conflict create drama; opposites in accord appear as beauty.

ACOUSTICS AND OVERTONES
from one note to seven and beyond

Any sound that can be perceived as a pitch or tone will have some periodicity in it, vibrating at a regular frequency with a specific mixture of overtone amplitudes (*see opposite*), creating a distinctive timbre. An oboe, sitar, or piano can all play the same tone, yet sound different. The vowels A, E, I, O and U are created by the trapping or releasing of overtones with the shape of the mouth and lips.

The other component of sound, noise, has no periodicity—a hammer striking, a finger plucking, a bow scraping, the sound on a television with no signal. Bands of noise are named by colour (white noise, pink noise, gray noise), and are part of the musical sounds an instrument can produce. The noise component of a sound can be compared to the consonants in language, with drums as plosives, shakers as fricatives, and cymbals as sibilants.

Essentially, musical sound can be described much in the way language sound can: a combination of tones that vary overtone content with a noise component that initiates the sound, sometimes continues it, and occasionally also closes it, the function of consonants, with an organizing rhythm and form.

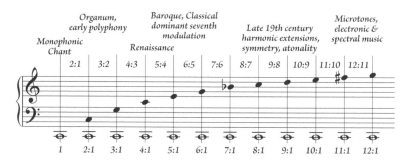

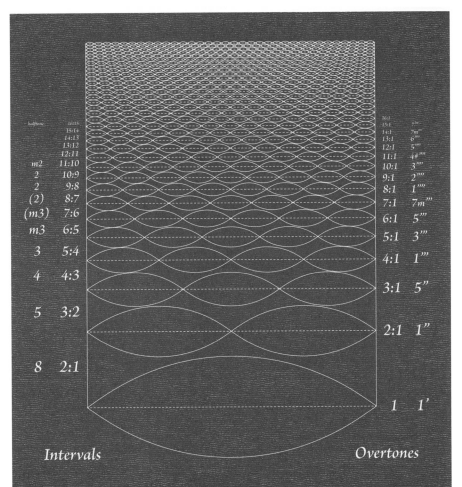

halftone:			16:1	
	16:15		15:1	
	15:14		14:1	7m''''
	14:13		13:1	6''''
	13:12		12:1	5''''
	12:11		11:1	4#''''
m2	11:10		10:1	3''''
2	10:9		9:1	2''''
2	9:8		8:1	1''''
(2)	8:7		7:1	7m'''
(m3)	7:6		6:1	5'''
m3	6:5		5:1	3'''
3	5:4		4:1	1''''
4	4:3		3:1	5''
5	3:2		2:1	1''
8	2:1		1	1'

Intervals

Overtones

Above: As a string vibrates, or tube fills with air, proportional waveforms are created in whole-number ratios that occupy the same space, giving rise to the harmonic series or overtones. Unity subdivides into infinitely smaller units. Each one of these overtones is a station or stopping point, a gravitational pole acting upon other tones nearby.

Left: The history of music in the West can be seen as a parallel to the overtone series, ascending upward and incorporating more and more of the series into harmonic thought. The relative distances of the intervals also suggest the time spent exploring those intervals, a journey from iconic objectivity to interior complexity.

UNDERSTANDING SCALES
streets and stairways

A scale is a collection of discrete tones that are a subset of the pitch continuum and that normally climb an octave in a certain number of steps, often seven. Most of the unique and beautifully diverse musical scales from around the world owe a large part of their heritage to the overtone series and use the fifth, the first tuned note, as the fundamental unit. Fifths are piled up, one atop another, and then transposed back down to a single octave. Thirds can also be prioritized to derive a scale (e.g. mean tone tuning) and a myriad of other methods all sculpt the different tuning systems that have emerged—each of them trying to solve the problem of locking a fluid, infinite curve or spiral into a grid or circle. The scale becomes a playground for a melodic drama unfolding the relative tensions of these overtones with the tones between them.

The basic stations are: 1-3-5-1, the major chord consisting of root, third, and fifth, created by the overtones 2:1 (the octave, the only note that when reached gives the distinctive impression of the fundamental tone below it, the same, yet different), then from 3:1, the fifth, which has the next quality of sameness, though it is in fact a different pitch entirely. Then 4:1, another octave, then 5:1, which becomes the third, generally conveying the major or minor quality of a chord, scale, or melody. Many 5-note and 7-note scales utilize this underlying structure, and in many variations, but the fundamental structures are 1-2-3-5-6 (pentatonic) and 1-2-3-4-5-6-7 (the major and various other scales). Seven is born from five.

A 7-note scale in a 12-note environment means that 5 notes will always be missing. In Middle-Eastern systems 7 notes are chosen (in performance) from 17; in India 7 are chosen from 22.

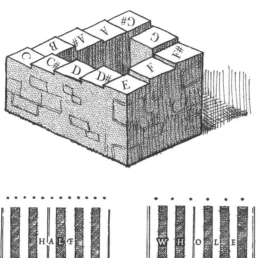

Left: This modified Penrose staircase shows the paradox of the musical octave. As we traverse a scale either ascending or descending, there is a simultaneous departure and return; we are coming and going at the same time.

Half steps: In chromatic tuning, e.g. on a keyboard, the distance from any note to its neighbour, white or black, is a half step or halftone. Western chromatic tuning has twelve equal half steps to the octave: C-C♯-D-D♯-E-F-F♯-G-A♭-A-B♭-B-C

Whole steps: Whole steps or tones are equivalent to two half steps. They too can traverse white or black, depending upon where they fall on the keyboard. The pair of mutually exclusive scales made solely of whole steps are: C-D-E-F♯-G♯-B♭ and D♭-E♭-F-G-A-C♭

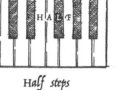

Half steps

Whole steps

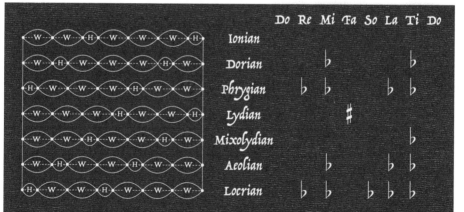

Above: The seven classical modes with their Greek names. Left: By sliding the starting point of the scale set one note to the right, all whole steps and half steps move to the left, creating six further modes from a seven note scale. Right: A way to generate the modes compared to the major scale (Ionian) by making a few chromatic adjustments, flat or sharp.

MEET THE INTERVALS
and the circle of fifths

A musical interval is the distance between two tones, and although tuned slightly differently from culture to culture, the same intervals are broadly found all over the world. Intervals can be thought of in two ways: the first is as an ever-contracting series of simple frequency ratios, so that the first interval (the *octave*) is a 2:1 relationship, the second (the *fifth*) is 3:2, and the third (the *fourth*) is 4:3. Then follow the *major third* of 5:4, the *minor third* of 6:5, the *second* of 9:8 or 10:9, and yet smaller intervals, with names like the *quarter tone*, *shruti*, *li*, *comma*, *apotome*, and *microtone*, depending upon the era and culture.

The second way of looking at this series is to compare intervals to the fundamental to which they all relate. This approach results in an octave (2:1), a fifth (3:1), another octave (4:1), a major third (5:1), another fifth (6:1), a seventh (7:1), another octave (8:1), a second (9:1), a third (10:1), and a *tritone* (11:1), etc. If the first view is relative, with each partial compared to its nearest neighbour, then the second view is absolute, as intervals take an absolute value compared to the fundamental. Both views are useful when taking an analytical approach to the construction of musical scales and the melodies that ultimately derive from them.

Notice the octave, fifth, and major third appearing in both systems. All scales around the world broadly contain these intervals in some form, with their precise tuning revealing slight variations and nuances in instrument construction and cultural tastes. Those using the first approach tune their instruments and derive their scales by the relationship of each overtone to one another, while others using the second approach relate intervals to their fundamental.

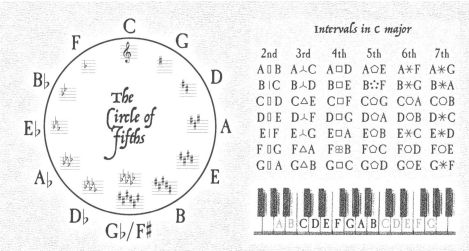

Intervals in C major

	2nd	3rd	4th	5th	6th	7th
	A▯B	A⅄C	A▢D	A⬡E	A✳F	A✲G
	B❘C	B⅄D	B▢E	B∷F	B✳G	B✲A
	C▯D	C△E	C▢F	C⬡G	C⬡A	C⬡B
	D▯E	D⅄F	D▢G	D⬡A	D⬡B	D✲C
	E❘F	E⅄G	E▢A	E⬡B	E✳C	E✲D
	F▯G	F△A	F⊞B	F⬡C	F⬡D	F⬡E
	G▯A	G△B	G▢C	G⬡D	G⬡E	G✲F

Above left: The circle of fifths is a diagram of common tones. Chromatic notes (sharps and flats) are added in the order of perfect fifths to preserve the whole and half step relationships in the major scale built upon any note. Any consecutive seven fifths on the circle yield all the notes in any one of the twelve major scales. Accidentals occur whenever an alteration to a scale must be made to preserve the syntax of the alphabet. Opposing notes on the wheel form a tritone, the symmetrical interval that cuts the octave in half.
Above right: A table of the intervals present in the key of C major, the white notes on a piano.

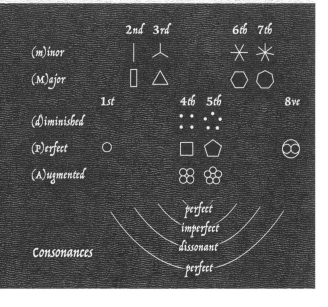

Right: A symbolic system of glyphs used in this book for describing musical intervals and their relative sizes. Note the shapes expanding and contracting with the size of the interval. Minor and diminished intervals are smaller by one halftone than their respective major and Perfect counterparts. Augmented intervals are one halftone larger than perfect. Major and minor are sometimes referred to as hard and soft (dur and moll).

The lower part of the diagram shows the symmetrical nature of the consonances of the intervals. Unison, the fourth, fifth and octave are perfect. Thirds and sixths are imperfect consonances, while seconds and sevenths are dissonances.

(m)inor

(M)ajor

(d)iminished

(P)erfect

(A)ugmented

Consonances

perfect
imperfect
dissonant
perfect

BASIC RHYTHMS
meter and the big beat

Rhythm is the component of music that punctuates time, carrying us from one beat to the next, and it subdivides into simple ratios just like pitch. Even in seemingly complex rhythms an underlying structure based on groupings of divisions into 2 and 3 is often perceptible. The march and the waltz are thus nodes in the subdivision of rhythm, and the tensions created by *polyrhythms* and *syncopation* push and pull against the gravity of these nodes, just as individual musical notes do in a scale. All of this happens through time, creating a framework of epigrams, disclosing their plight or journey, existing within a system of rules.

Rhythmic structures are organized into *measures* for the purpose of notation, which denote time parceled into groups of beats. In 4/4, each measure has four beats marked by quarter notes, which often show up in groups of four measures. Within most rhythms a pulse of strong and weak beats, or strong and weak parts of beats, also exists, and *chords* are placed in each measure at either the *anacrusis* or the *ictus*, 'between' or 'upon' the beats, to convey harmonic movement and reinforce the sense of tonality. The unfolding and varying of the resulting tensions and releases through time is responsible for much of the emotive and expressive power of rhythm.

The rate at which events pass is also a crucial component of any rhythmic texture, often measured by beats per minute (BPM). A pulse's subdivisions are partly meaningless without knowing its rate or *tempo*.

bpm: 40   *largo*   60   *adagio*   76   *andante*   108   *moderato*   120   *allegro*   168   *presto*   200

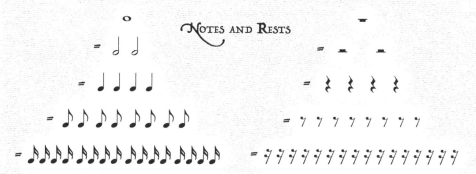

Notes and Rests

Above: The basic duple subdivisions of the beat as notes (left) and rests (right). The values are, from top to bottom, whole, half, quarter, eighth, and sixteenth. This process can continue up to one-hundred-twenty-eighth notes. The rest occupies the same potential space as a note does, and provides breath and space, helping to clarify different parts.

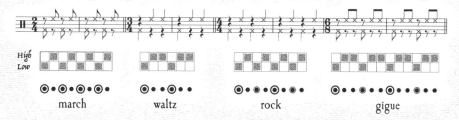

High
Low

march   waltz   rock   gigue

Musical styles are often expressed by their rhythmic patterns, and the grouping and subgrouping of their component strong and weak beats. Shown here are some of the most basic ones, built upon 2 and 3, simple and compound.

In a time signature, the numerator indicates the number of beats per measure, and the denominator indicates which subdivision will receive the beat. The measure, or bar, separated by lines, indicates a rhythmic cell, one repetition of the basic grouping or cycle.

Long and Short

Duples	Triples
Iamb ∣ —	Tribrach ∣ ∣ ∣
Trochee — ∣	Dactyl — ∣ ∣
Spondee — —	Amphibrach ∣ — ∣
Dibrach ∣ ∣	Anapest ∣ ∣ —
	Bacchius ∣ — —
	Antibacchius — — ∣
	Amphimacer — ∣ —
	Molossus — — —

The poetic feet measure all the possible combinations of long and short in duple and triple groupings, twelve total.

TONE TENDENCIES
tension and release

Because of the powerful gravity and stability of the stations of the scale, notes that deviate from these are perceived as transitional. Whole and half steps, and even minor and major thirds, all manifest varying degrees of tension, which is then released when a station is reached, or when transitional notes are traded for less dynamic transitional notes. Additionally, notes that are farther from a stopping point or station are less active or dramatic in relation to it than adjacent ones. Minor scales have basically the same set of tensions as major scales (*see page 274*).

A further level of complexity occurs in music that has chord progressions and modulations, as the set of tensions can change. Here the initial root, third, and fifth of a chord (and scale) are the stations, with the rest of the tones intermediary, but as the chord changes, the root, third, and fifth of the new chord become the new stations, or secondary stations. However, without an actual modulation, the importance of the primary set of relations is not lost in memory, and a tiered set of relationships is created. Since chords can be constructed with any note of the scale as a root, they can both take on the same stable or transitional aspects as the roots upon which they are built already possess, and contain their own subset of stations, forming two tiers of stability or instability working at any given moment. The genius of a good melody involves the understanding of these two tiers of relationships, and the skillful implementation of manipulating expectation and result against those natural tensions, based largely upon the utilization of memory, incorporating expectation and fulfillment.

Western	**Do**	**Re**	**Mi**	**Fa**	**So**	**La**	**Ti**	**Do**
	ROOT	ROUSING	STEADY	DESOLATE	GRAND	SAD	LEADING TONE	
	FIRM	HOPEFUL	CALM	AWE	BRIGHT	WEEPING	PIERCING	
Indian	**Sa**	**Re**	**Ga**	**Ma**	**Pa**	**Da**	**Ni**	**Sa**
	BASIS	JOY	ANGER	CLEANSING	SPEECH	NIGHT	AGITATED	
	STRONG	SEXUALITY	WRATH	YIN	YANG	PLEASURE	EFFORT	

Above: The seven tones of the major scale, their Curwen hand signs, and associated qualities. Horizontal hand signs indicate the stations, the others signify the tendencies of the intermediary tones, pushing down or pulling up.

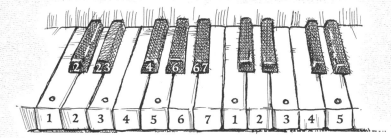

Above: The main theme of the 2nd movement of Beethoven's Pathétique Sonata in A♭ Major with Kodaly hand signs. The accidentals (E-natural in bar 5 and A-natural in bar 7) act as temporary leading tones, so they receive the 'ti' hand.

Above: Intermediary tones each come in varying flavours. The second can be either lowered, natural, or raised; the fourth can only be natural or raised. If it were lowered it would become the third. The sixth can come in three flavours, but the seventh is also limited, for it can only be lowered or natural. If it were to be raised it would become the octave.

255

BASIC HARMONIES
triangles and triads

The major triad, which occurs naturally in the harmonic series as a pair of thirds (a major, then a minor, adding to a fifth), is the foundation of tertial music (chords constructed in thirds) around the world, the perfect fifth and major third being, after the octave, the most stable and resonant intervals, derived from the overtones.

Moving a note from the bottom to the top of a triad creates an inversion (*below*), with the same notes, but with a new bass. Notice how major triads in first inversion have two minor intervals, giving them an opposite flavour. The same holds true of minor chords, which in their first inversion sound markedly major, since two of their three intervals are major. Diminished and augmented chords are often said to be rootless, as they have no stable fourth or fifth.

Inversions conspire to strengthen or weaken the importance of the root. In root position the fundamental intervals are all in place as in the overtone series, the bottom note receiving the identity of the chord built upon it. In the first inversion, the third of the chord is in the bass but has no strong intervals above it to emphasize its importance. Instead, the root, now at the top, is supported by a perfect fourth just below it, another strong architectural interval. The same is true with the fifth in the bass, the second inversion, where a perfect fourth again supports the root. The combined notes thus always point to their root position, stacked in thirds.

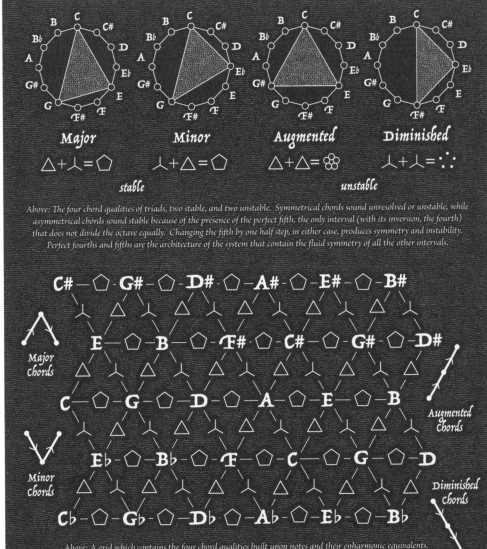

Major   **Minor**   **Augmented**   **Diminished**

△ + ⅄ = ⬠   ⅄ + △ = ⬠   △ + △ = ⬡⬡   ⅄ + ⅄ = ∴

stable   *unstable*

Above: The four chord qualities of triads, two stable, and two unstable. Symmetrical chords sound unresolved or unstable, while asymmetrical chords sound stable because of the presence of the perfect fifth, the only interval (with its inversion, the fourth) that does not divide the octave equally. Changing the fifth by one half step, in either case, produces symmetry and instability. Perfect fourths and fifths are the architecture of the system that contain the fluid symmetry of all the other intervals.

Above: A grid which contains the four chord qualities built upon notes and their enharmonic equivalents. Follow the keys on the left and right to spell any of the four chords in thirds.

257

BASIC MELODY
steps and leaps, contour and gesture

A melody is created by the succession of tones through time. Step by step, note by note, an outline is formed, a path carved. Gestures appear, like the inflections used in speech, or the dialectic of rising and falling tones, or the contrast of high and low notes. A distant leap feels large and grandiose, a small one more fluid and gentle. Curved or jagged contours can be suggested.

Melodies are normally a mixture of small *steps* and larger *leaps*, with a leap in one direction inducing a yearning for completion by a step in the opposite direction, leaving a gap to be filled in. The continuous nature of melody means that when notes stray far, the listener, following the path to find out where it leads, likes them to remain connected and return. This is often manifested by a rhythmic intertwining of tones in and out of the stations of the scale.

The expressivity of a melody comes in part by the tension and release of the intermediary notes of the scale, their rhythmic placement on a strong or weak beat intensifying or diminishing their effect. Sometimes a melody can act as two melodies, by leaping up and down, thus alternately maintaining two independent threads, each on their own pitch level (or *register*). Other melodies rely on pitches predominantly rising or falling for their effect.

Melodies in vocal music are either *melismatic*, with many pitches to one syllable, or *syllabic*, with one pitch per syllable.

Silent pauses, or rests (*caesura*) are essential to melodies. They allow time for breath, reflection, and interaction with the music. Listeners wait for the next event, suspended, anticipating. A well-placed rest in a theme can be a powerful musical moment.

Above: Melodies are easily expressed visually. They weave between the strong stations of the scale and the weaker, transitional parts, which contribute to the tensionality and expressive shape of any melody.

Upper Neighbour
Decorates a station from above, weak beat

Lower Neighbour
Decorates a station from below, weak beat

Passing Tone
Acts as a bridge between two stations, weak beat

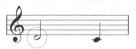

Appoggiatura
A non-station event that leans on a station, strong beat

Escape Tone (échappée)
A non-station note that escapes a step in one direction and leaps in the other

Cambiata
A neighbor group, upper and lower, often dissonances decorating a station

Non-station notes can be either accented or unaccented, falling on strong or weak beats (except the appoggiatura, which is always accented). When they occur on strong beats, they compete more strongly with the stations to convey the harmony.

Trill   Trill   Mordent   Turn   Grace note

Above: Melodic ornaments and embellishments, such as (from left to right) trills, mordents, turns, and grace notes, are also melodic formulae, on the smallest scale.

CHORD PROGRESSIONS
tonic, dominant, and subdominant

Chords take on the identity of the station upon which they are built, reinforced by the tone tendencies, so that as the chords move in a progression or succession, they push and pull on their neighbours, reinforcing the key. As notes in a chord collaborate to emphasize one pitch, all chords collaborate to strengthen or weaken the tonic.

The strongest progressive motion a root can have is a fifth, either downward or upward, highlighting the close kinship and relative gravity of three consecutive pitches on the circle of fifths. The most basic chord movement then is *tonic* (I) to *subdominant* (IV) to *dominant* (V) and back to tonic (V), although other triads in the scale can substitute for these three basic functions without sacrificing wholly the function or temporal meaning of their placement (*below*). I and *iii* share 3 and 5, so can substitute for each other. Likewise, *ii* and IV share two tones, as do *vi* and I, and V and *vii°*. The more common tones, the smoother and more gradual the harmonic motion.

There are essentially three states in tonal harmonic progression: starting, departing, and returning, which are repeated and cycled to reinforce the tonic. Chords with more symmetrical intervals (thirds) are unstable or require resolving, and have a dominant function, while harmonies having perfect fourths and fifths (the only asymmetric intervals) are more restful and are non-dominant.

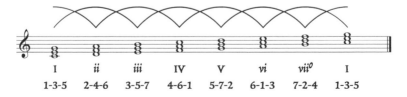

I	*ii*	*iii*	IV	V	*vi*	*vii°*	I
1-3-5	2-4-6	3-5-7	4-6-1	5-7-2	6-1-3	7-2-4	1-3-5

DOMINANT *(active)*   ☉   5   V

MEDIANT   ○   3   iii

SUPERTONIC   ○   2   ii

TONIC (HOME)   🜨   1   I

SUBTONIC *(leading tone)*   ○   7   vii°

SUBMEDIANT   ○   6   vi

SUBDOMINANT *(passive)*   ☾   4   IV

rising / *falling*

Left: The seven roots of the the seven triads each have a specific spatial relationship to the tonic. The tonic/home (Earth), has great gravity in tonal music, strongly pulling the two bodies closest to it (ii and vii°). The Sun and Moon, a fifth away up or down, have the weakest gravity and can exist as separate, yet related entities.

V negates I, while implying its presence by resolving to it. IV already contains the root of I in its triad, so is more closely related to the tonic than V.

The mediant (middle third) and submediant (lower middle third) have moderate gravity and are weaker stopping points, sharing two tones each with two chords.

ROOT MOVEMENTS

2ND — Root moves up or down by a second. Strong movement, no common tones.

3RD — Root moves up or down by a third. Gradual movement, two common.

4TH or 5TH — Strongest movement, around the circle of fifths. One common tone.

PACHELBEL'S CANON   I – V – vi – iii – IV – I – ii – V

HEART AND SOUL   I – vi – ii – V

LET IT BE   I – V – vi – IV – I – V – IV – I

progression ⟷ *regression*

Above: Upward root movement is progressive. Downward movement tends to be more regressive and relaxed. Most chord progressions are a circular balance of up and down, near and far. Apart from the special relationship between I, IV, and V, the fewer common tones between two chords, the greater the sense of movement.

Primary **MAJOR**	I	IV	V	I
Secondary **MINOR**		vi	iii	
	(iii)	ii	vii°	(vi)

SUBSTITUTIONS

Left: When reducing secondary chords to their primary function, we can see whether a chord succession is progressing or regressing, active or passive, and thereby gauge the energy and rate of movement.

INSTRUMENTATION
the textures of timbre

Musicologists identify six basic types of musical instrument: membranophones (membranes), chordophones (strings), idiophones (struck), metallophones (metallic), aerophones (air), and electrophones (electronic).

Wind instruments generally have an open end, and often a conical shape to release sound into the air. Wind can be moved through a narrow space to vibrate a reed, or two, or sound can be generated by the buzzing of lips (blowing through a tube), as with brass instruments. Strings can be either plucked or bowed, and can resonate in sympathetic accord with other strings. Percussion instruments move air quickly, abruptly, and noisily, and have membranes and means of striking them. They provide contrast to the smooth, sustained tones of melody and harmony, punctuating with shakes, sizzles, tingles, and rings. All cultures utilize percussion in their music, and many brass and stringed instruments reveal the algorithmic spiral or curvature of pitch. Some cultures also believe instruments contain animal souls that sing when the gut or skin vibrates. Many instruments resemble the structures of the ear, both being part of the vibration duality, instruments generating and the ear receiving. The voice can imitate most (*the basic ranges of male and female voices are shown below*).

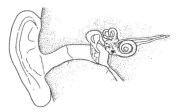

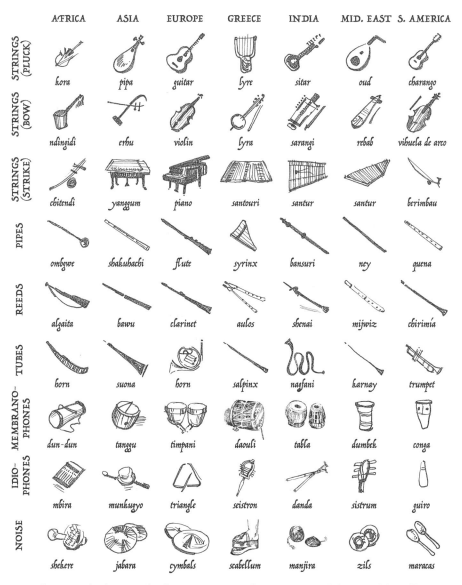

	AFRICA	ASIA	EUROPE	GREECE	INDIA	MID. EAST	S. AMERICA
STRINGS (PLUCK)	kora	pipa	guitar	lyre	sitar	oud	charango
STRINGS (BOW)	ndingidi	erhu	violin	lyra	sarangi	rebab	vihuela de arco
STRINGS (STRIKE)	chitendi	yanggum	piano	santouri	santur	santur	berimbau
PIPES	ombgwe	shakuhachi	flute	syrinx	bansuri	ney	quena
REEDS	algaita	bawu	clarinet	aulos	shenai	mijwiz	chirimia
TUBES	horn	suona	horn	salpinx	nagfani	karnay	trumpet
MEMBRANO- PHONES	dun-dun	tanggu	timpani	daouli	tabla	dumbek	conga
IDIO- PHONES	mbira	munkugyo	triangle	seistron	danda	sistrum	guiro
NOISE	shekere	jabara	cymbals	scabellum	manjira	zils	maracas

Above: Examples of string, wind, and percussion instruments from various times and places around the world.

More Complex Rhythms
dynamics, articulation, elocution, and syncopation

Sounds evolve over time, and *envelopes* (*below*) use three basic phases (inception, continuation, and closure) to characterize different qualities of volume over time. *Staccato* notes have a short and detached quality, *tenuto* notes are slightly lengthened to connect them (in *legato* fashion) to those nearby, while an *accent* indicates a strong start to a tone, emphasizing its initiation.

Piano and *forte* are soft and strong indications used to suggest volume or amplitude. The pianoforte (piano) was so named for its ability to play both loud and soft, in contrast to earlier keyboard instruments, the harpsichord and clavichord, which could not.

Articulations and dynamics in music notation shape passages and contribute to the sense of character and mood, whether playful, doleful, whimsical, or aggressive. Articulations affect the presence of noise in the instrument, acting as consonants upon vowels.

Opposite are shown dotted rhythms, used in triple notation. Notice the dot to the right of the note, rather than above or below it (as with staccato). Rhythms continue to complexify and subdivide in syncopation and polyrhythms (*see too page 396*).

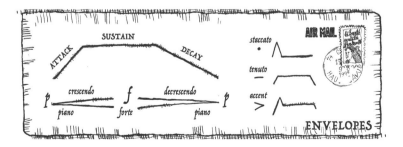

DOTTED
NOTES AND RESTS

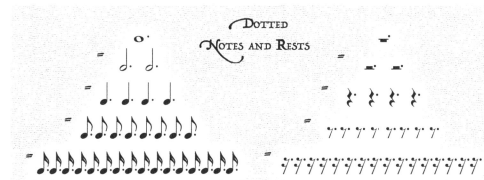

Above: Dotted notes (left) and rests (right). Dotted rhythms are equal to one and a half times the value of the undotted note or rest, and are useful for obtaining triple subdivisions.

Above: A sample bass line in 3/4, illustrating the relationship of three quarters to a dotted half note, both occupying the same amount of time.

Above: In 6/8, three eighth notes are equivalent to two dotted eights, which in turn equal one dotted half.

Right: Tuplets can represent any number of subdivisions, depending upon the metric context in which they appear. Tuplet subdivisions are effectively identical to the subdivisions of frequency in the overtone series. By this point in the notational system, virtually any subdivision of the beat can be notated.

POLYRHYTHMS

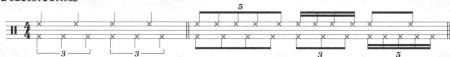

Above: Polyrhythms occur whenever different subdivisions of the beat appear together, such as 2 against 3, 5 against 4, 4 against 3, 2 against 5, and so on.

SYNCOPATION

Above: Syncopation is the shifting of strong beats on to weak beats, thus changing our sense of predictability and steadiness. Too much syncopation and we are lost for a pulse. Not enough, and we soon grow bored. Listening to how different cultures push and pull against their metric structures can be tracked largely through the use of syncopation.

FORM AND STRUCTURE
where am I going and how did I get here?

Musical structure tends to unfold in parts or sections. An idea, mood, or motif is first presented before something arrives that changes or contrasts it, while nevertheless relating to it, creating a sense of unity and, ultimately, arrival or return. This unfolding pattern also helps orient the listener in time, so that, using their attention and memory, they can tell where they are in the musical texture. Without it they would be adrift in a sea of unrelated ideas, and some music is intentionally composed this way for that very effect.

Most people go through vicissitudes of emotion in their life, leaving home, going out into the world, having adventures, and ultimately returning home. A life's journey is like a musical composition, born into the world from nothing, living for a time in form and structure, dancing spontaneously on the edge of chaos and order, and then finally returning. In this respect Western music tends to be more linear, Eastern music more cyclical.

Musical time can be visualized as a storyboard (*below*), each segment expressing the essence of a particular character, meaning, intention, and purpose of a section or movement. Frequently these sections are ordered with consideration for the attention span of the listener, in varying degrees of complexity and engagement, like a ceremony involving invocation, meditation, and dance.

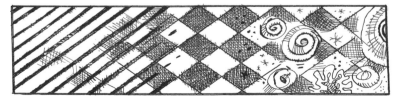

ʙᴀsɪᴄ ᴍᴏʟᴇᴄᴜʟᴀʀ ꜰᴏʀᴍs ᴏꜰ ᴡᴇsᴛᴇʀɴ ᴍᴜsɪᴄ

BINARY	ROUNDED BINARY	TERNARY

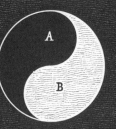

Above: Binary forms are simply two sections that contrast each other to make one whole, AB.

Above: Rounded binary brings back a truncated version of the A, usually just enough to suggest or remind the listener of where they've been, 'ABA'.

Above: Ternary forms present the A in recapitulation, yielding an ABA structure.

ᴄᴏsᴍᴏʟᴏɢɪᴄᴀʟ ᴅɪᴀɢʀᴀᴍs,
some ancient keys to nuances in the evolution of form - showing how time, space, and manifestation interrelate

Above: The humours, seasons and elements of antiquity. These correspondences illustrate a kind of interrelated dynamism and flow to time, character, personality type, and the human experience.

Above: The Far-Eastern system of elements, demonstrating principles of balance and imbalance, and showing how the flow of time and the elements correspond to interact and influence each other.

MORE COMPLEX HARMONIES
sevenths and their inversions add suspense

Continuing to stack thirds beyond the triad yields major, minor, or diminished sevenths, with their dynamic pull toward the tonic. As the root and fifth of a chord provide architectural structure, the third and seventh provide feeling and flavour, push and pull. Sometimes referred to as *guide tones*, they lead the movement of harmony from chord to chord, adding to the intensity of the forward drive. The presence of two guide tones in a chord maximizes this, e.g. the dominant 7th, whose third and seventh often resolve to the tonic root and third.

Hollow *suspended* chords, where the third is replaced with its junior or senior note, are shown below. Similarly *add* chords take on a colour without compromising any of the three basic notes of the triad, the 2, 4, and 6 sweetening the overall sonority.

If the lowest tone is kept constant, as chords move around it, then we are in the presence of a *pedal point*, so named because of the ability of the organ to sustain bass tones played with the feet while changing harmonies played with the hands on the keyboard, keeping a central bottom tone in place, which may or may not be the root of the chord. In fact, a pedal point doesn't have to belong to the chord at all, differentiating it from an inversion.

268

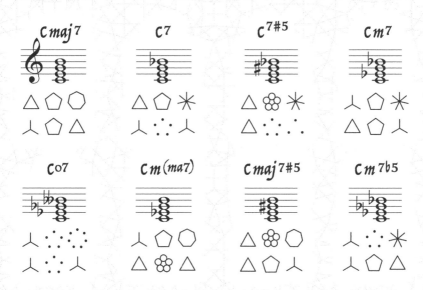

Above: The eight basic seventh chords derived from stacking thirds. Note the interval content of the major chords versus the minor chords, also the open and expansive quality to chords with more perfect and major intervals, and the more closed and contractive quality to chords with diminished and minor intervals. The maj7, m(ma7) and maj7#5 have the most large intervals. The 7#5 and m7 have an equal balance, and the 7, °7, and m7♭5 (also known as half-diminished ø) are made up mostly of smaller intervals.

		C maj7	C7	C7#5	C m7	C o7	C m(ma7)	C maj7#5	C m7♭5
3RD INV	6 4 2	B	B♭	B♭	B♭	B♭♭	B	B	B♭
2ND INV	6 4 3	G	G	G#	G♭	G♭	G	G#	G♭
1ST INV	6 5 3	E	E	E	E♭	E♭	E♭	E	E♭
ROOT	7 5 3	C	C	C	C	C	C	C	C

Above: Inverted sevenths. The figured bass notation (second column) indicates the intervallic placement of notes above the bass, which in fact indicates the inversion, though the 1, 3, and 5 are assumed and not always written. We still hear the identity and therefore function of the chord as though it were in root position. The brain reassembles the notes into their closest formation, regardless of how the individual notes are voiced, even when spaced openly across octaves (see too page 256).

TONALITY AND MODULATION

there's no place like home

Tonality, or the sense of being in a particular key, is most easily created by sounding the I-IV-V-I pattern (*see pages 260–261*).

The tonic chord has the same function as the tonic note of the scale. It is the place where things begin and end, and to which all things relate. In tonal music the leading tone is always used to point to the tonic, and, as with the circularity of the scale, all other stations of the scale and their chords serve either to strengthen or weaken the relative gravity of the tonic, departing or returning.

The mighty tonic, however, can be destabilized. Chords other than the tonic can be strengthened by the introduction of their respective leading tones (the third of any dominant chord, a half step below a root). Notes outside of the key can make an appearance to point to other roots as possible tonics. The sense of a second key can emerge. If this happens within the appropriate time, and with the repetition of the I-IV-V-I chords of the new key, then a full *modulation* occurs, and a new tonic is formed. Without these affirmations, the movement is temporary, and only a *tonicization* has occurred.

After a time, the ear may become accustomed to the new key, but tonal music often returns to the first by reinstating the changed note, creating the sense that we never really left. Before long, a relationship of keys emerges, reflecting the relationships of chords, which in turn reflect the relationships of the individual tones in the scale.

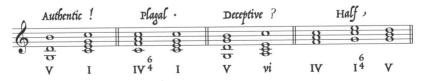

Right: In this basic I-IV-V-I progression, note how the tritone, or diminished fifth formed between the unstable leading tone B and upper F of G7, contracts inward to the stable root and third, joined by the upward movement of the dominant in the bass. This maximizes the finality and reality of the chord in the I position being the true home.

Left: Cadences (endings). A cadence in music works much the same way as punctuation in language. The authentic cadence has a strong degree of finality due to its leading tone and its root resolving to the tonic. The plagal cadence is subtler, lacking the active drive of the dominant, the 'A-men' in much European sacred music. A deceptive cadence resolves someplace other than the expected resolution, the tonic, often a surprise. The half cadence is open ended, ending on the dominant, sometimes with an intervening tonic chord in 2nd inversion to heighten the expectation of resolution.

I   IV   V7   I

To Modulate from the key of X to the key of Y
find a pivot chord that is in both keys and substitute for IV

So On the wheel above, moving clockwise:

Moving 1 click: vi in X is ii in Y OR iii in X is vi in Y

Moving 2 clicks: iii in X is ii in Y

(Moving 3 clicks or more requires borrowing from the parallel minor)

In the key of X: 1 - Pivot - V - 1

In X: 1 - Pivot

In Y: Pivot - V - 1

Example: to change key from G to A and back (use Bm):

G - Bm - D7 - G, G - Bm - E7 - A

A - Bm - E7 - A, A - Bm - D7 - G

MODAL, TONAL, DRONAL
world systems and scales

In every musical scale there is a much-used primary set of tones, and a smaller secondary set, used to colour the first. Basic primary pentatonic, five-note scales form the backbones of many scales around the world (*below*), the most simple deriving from four consecutive fifths. Other scales follow the gravitational forces of the stations, and use chromatic alteration to communicate tension and emotion. Scales with more half steps can be tighter, filled with introversion, chromatic complexity, and pathos, while more diatonic scales can be extroverted, simple, affirming, and expansive.

There are essentially three kinds of pitch organization. *Modal* music, which does not modulate, uses leading tones liberally, sometimes not at all; harmonic movement is possible, and chords can be borrowed from other scales. In *tonal* music, which does modulate, the five secondary tones not used in the major scale have a relatively fixed relationship to the primary set, determined by the circle of fifths, ♯ 4, ♭ 7, ♯ 5, ♯ 1, ♭ 3, each temporarily suggesting the scale to which the secondary tones act as leading tones. These accidentals only last for the measure in which they occur, and are restored in subsequent measures, sometimes followed by a courtesy accidental as a reminder. *Dronal* music has no harmony, the scale itself being the harmonic universe, so the intervals all relate solely to the still point, the drone, with a full chromatic range available.

MODAL

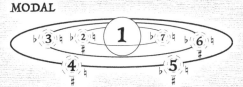

In modal music, the dominant chord need not be major, or even used, and leading tones can occur in positions other than in the standard major/minor. With good harmonic variability and softened tone-tendencies, this is the most common type of harmonic system.

"The Handsome Cabin Boy"

TONAL

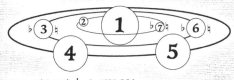

Tonal music involves just two modalities, major and minor. The positions of the half steps are highly organized. The dominant chord is always major, pointing to the tonic with its powerful leading tone (the third of its chord, also the 7th of the scale).

Mozart Sonata in A, KV 331

DRONAL

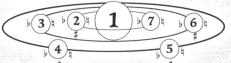

Dronal music, melodic rather than harmonic in nature, has all the notes of the scale assuming variable roles, taking on importance by interval and repetition, all of them always relating to the central point, resting tone, or drone.

Raga Shivranjani

THE THREE MINORS
natural, harmonic, and melodic

The active role of leading tones in harmonic music creates complexities for minor keys. The natural dominant chord in minor is not a major chord, and the need for a leading tone necessitates altering it, and the scale from which it is built.

In the case of *natural minor*, we are in the presence of the Aeolian mode, which occurs naturally as the relative to any major scale, started a third lower. Raising the third in the dominant chord, the seventh of the scale, creates *harmonic minor*, altered for harmonic purposes. The scale that results from this, however, contains an audible gap between the flatted minor 6th and the natural major 7th, an augmented second, which does not always work melodically, sounding like an interval from non-Western music. To smooth out this melodic gap, the 6th scale degree is also altered, raised, so that the ascent in minor resembles a major scale in its upper four notes. This is *melodic minor*, which is sometimes said to have two forms: ascending and descending, raising and lowering the 6th and 7th accordingly. In fact, descending melodic minor is identical to natural minor. Triads of the three minors are shown below.

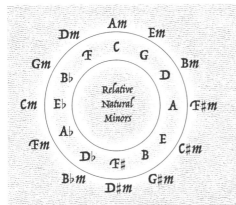

Opposite page: These three minor scales are used interchangeably depending upon the composer's needs; and adhere to basic rules about the use of chromaticism in the minor mode. These scales, like major/Ionian, can be used as starting points to generate others, so in fact there are seven modes of harmonic minor and seven modes of melodic minor (shown below).

Left: Natural minor is identical to a major scale a minor third above. So A minor and C major are in fact the same notes, shifted a third apart. Natural minor, being Aeolian, already shares the same pitches as its relative major, Ionian, therefore they share the same key signature (see page 251).

The 7 modes of Harmonic minor

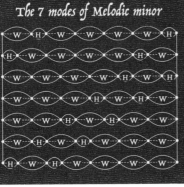

The 7 modes of Melodic minor

MORE INTERVALS
the big get bigger and the small get smaller

Just as chords are invertible, so too are their constituent intervals. 3rds invert to become 6ths, and 7ths become 2nds. Inverted major intervals become minor, inverted diminished intervals become augmented, and vice versa. Inverted intervals have the same basic function as their non-inverted counterparts, but possess a greater sense of uncertainty. Composers play with them by further raising and lowering them. For example, extending a major 6th yields the interval of an augmented 6th, identical in sound to a minor 7th, but functioning quite differently. Remember that spelling counts, and that a minor 7th tends to resolve inward, while an augmented 6th tends to resolve outward—big gets bigger and small gets smaller. Similarly, contracting a minor 7th yields a diminished 7th, the same as a major 6th, but again functioning wholly differently. As a major 6th is likely to fall by a whole step to the 5th, or rise to the major 7th, a diminished 7th almost always falls by a half step. Spelling and syntax indicate behavior and directionality.

Shown below are the only three possible *octatonic* or diminished scales, since they are symmetrical. They fall into the repeating pattern of half-whole or whole-half. Also shown are the only two wholetone scales, built entirely of whole steps. They can evoke a mysterious sense of ambiguity and the unknown.

The three diminished scales
and
The two whole-tone scales

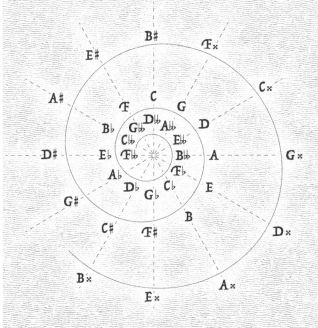

Left: Enharmonicism & Chromaticism. Combining the circle of fifths with the spiral of fifths produces all of the possible 35 spellings for the 12 notes. Remember there are 7 members of the scale, ABCDEFG, and 5 chroma (colours) for each. The outermost and innermost spellings are rarely used (B✗, E✗, F♭♭, C♭♭). Whenever a chromatic alteration must be made to a scale, an accidental is correspondingly used (a lowered flat is double flat, a raised flat is natural, a lowered sharp is natural, a raised natural is sharp, and a raised sharp is double sharp).

This diagram is of equivalence between spellings and assumes equal temperament. It should not be confused with a common diagram demonstrating tuning problems created by the Pythagorean Comma (covered in BOOK IV of this volume).

	2nd	3rd		6th	7th
(d)iminished			I N V E R S I O N S		
(m)inor					
(M)ajor					
(A)ugmented			I N V E R S I O N S		

Left: The imperfect consonances (3rds and 6ths) and dissonances (2nds and 7ths) occasionally come in two extreme forms beyond major and minor. Chromatically extending a major interval further turns it into an augmented interval. Likewise further contracting a minor interval makes it diminished, just as a perfect interval becomes diminished when contracted.

Complex and chromatically rich music sometimes will use these enhanced or stretched intervals to push and pull with even more energy against the overall tonal structure.

FURTHER MELODIC ELEMENTS
epigrammatic development

Humans are fundamentally pattern-based beings, and artists world-wide, visual and acoustic, have used this fact to manipulate audiences for centuries. An idea or epigram is articulated, before undergoing a series of transformations, being reinforced or denied. The unfolding of this drama of denial and acceptance is the narrativity of music, and is played out in three basic ways:

Repetition, or thesis. The easiest thing for an epigram to do is to assert itself, and this is accomplished by repetition. Repetitions are helpful because they are highly orienting for the audience in the context of a given narrative. They are the anchors of time.

Contrast, or antithesis. The drama begins. A new epigram is presented, possibly seeming to contradict the previous one, and creating the tension of a new set of opposites. A completely contrasting epigram, with no epigrammatic transference, is a denial.

Variation, or synthesis. A reconciliation of the two poles of repetition and contrast, sometimes viewed as a fulcrum (*below*).

Melodies, songs, and symphonies all use these three degrees of epigrammatic transference; from total (repetition), through partial or transformative (variation), to none (contrast). The mind absorbs the meaning of each new idea, comparing it to recent and distant events. Attempting to cognize the parts produces anticipation, which may be affirmed or denied. Narrative artists exploit this faculty to create their books and movies, melodies, and rhythms.

Species Counterpoint

1st species: Note against note. Only consonances are permitted. Careful avoidance of parallel leaps or steps, especially to a perfect interval, except at a cadence.

2nd species: Two notes against one. Passing tones make their appearance, as dissonances which are permitted only on weak beats. The contrapuntal line can start on a rest.

Third Species: Four (or three) notes against one. Passing tones, neighbour tones, and now escape tones can be used, still adhering to consonances on strong beats.

2nd species: Two notes against one. Passing tones make their appearance, as dissonances which are permitted only on weak beats. The contrapuntal line can start on a rest.

Fourth Species: Suspensions—offset notes. Consonance is prepared on a weak beat, and when the pitch in the cantus firmus changes, a dissonance is created by the sustained tone on a strong beat, and resolved by a step downward.

Above: Species Counterpoint, a system of rules for writing polyphonic music, dating back to the 16th century. The cantus firmus, or bottom line, is joined by a new melodic line on top. fifth species is the combination of the previous four, known as florid counterpoint. Counterpoint is the simultaneity of melodies, each line running independently of the others horizontally, and aligning to make harmonic sense vertically (often at strong parts of the meter, to convey the harmonic framework). The four-part texture of soprano, alto, tenor, bass in contrapuntal music later became melody, bass line, and 'inner voices'. The highest voice became the most melodically important, the others taking on a more supporting role, eventually becoming chords, with a collective identity of their own. This more homophonic texture is the 'melody and accompaniment' with which we are so familiar. There are three types of contrapuntal motion: parallel, contrary, and oblique.

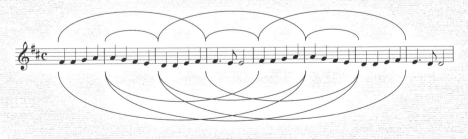

Above: Theme from Beethoven's 9th Symphony, showing symmetrical nature of epigrammatic melodic movements.

COMPLEX CHORD PROGRESSIONS
getting out of the box

To develop a richer harmonic palette and get out of the i-iv-v-i box, a more complex structure can be developed by borrowing chords from a parallel scale, thus facilitating longer excursions.

Because it is the root of a chord that imparts its functionality, we can freely substitute other chords built upon the same scale degree (*opposite top*) and still preserve the harmonic essence. So a major subdominant (iv) can be substituted for a minor one (*iv*), or a minor mediant (*iii*) can be replaced with a major mediant (♭ iii) and its inflections (♭ iii+). As long as basic cadences occasionally occur to reinforce a tonic, chords can be borrowed relatively freely.

The dominant seventh chord is well-suited for substitution because of its symmetrical tritone (*see below*). When the root shifts by a tritone, the 3rd and 7th of each chord exchange places. The spelling of this interval changes enharmonically to preserve the syntax, but the sound is the same. In fact, as we move around the circle of fifths with seventh chords, the 3rd and 7th of each chord exchange places and slip and slide by steps, often referred to as step progressions or guide tones, a reciprocity that maximizes the forward drive of harmonic motion. In most chords, it is the root, 3rd and 7th that are sufficient to communicate the harmonic function, so when voicing chords, the 5th can frequently be omitted, since it only reinforces the tonic structurally. If, however, the 5th is altered (♯ or ♭ , augmented or diminished), then its colour is included as well.

280

Scale Degree	1	2	3	4	5	6	7	8
Major	C	Dm	Em	F	G	Am	B°	C
	I	ii	iii	IV	V	vi	vii°	I
Natural Minor	Cm	D°	E♭	Fm	Gm	A♭	B♭	Cm
	i	ii°	♭III	iv	v	♭VI	♭VII	i
Harmonic Minor	Cm	D°	E♭+	Fm	G	A♭	B°	Cm
	i	ii°	♭III+	iv	V	♭VI	vii°	i
Melodic Minor	Cm	Dm	E♭+	F	G	A°	B°	Cm
	i	ii	♭III+	IV	V	vi°	vii°	i
Secondary Dominants	C7	D7	E7	F7	G7	A7	B7	C7
	V^7/IV	V^7/V	V^7/vi	$V^7/♭VII$		V^7/ii	V^7/iii	V^7/IV
Minor secondary Dominants			E♭7			A♭7	B♭7	
		$V^7/♭VI$				$V^7/♭II$	$V^7/♭III$	

Above: A table of substitutions. Secondary dominants act like the primary dominant in that they possess a tritone, and suggest a resolution a fourth up or fifth down. They require a chromatic alteration to the basic scale in use. The new leading tone that results temporarily suggests an alternate key or scale, but this is usually brief, either occurring as a passing chord, or as a tonicization. The presence of borrowed chords weakens the strength of the tonic, but often provides a lovely shading or colour, partly due to their violating our expectations about which chords we expect to hear in a given key.

Amazing Grace

I	V^7/IV	IV	I
I	I^6	V	V^7
I	V^7/IV	IV	I
I^6_4	V^7	IV^6_4	I

Greensleeves

i	♭VII	♭VI	V^7
i	♭VII	♭VI - V^7	i
♭III	♭VII	♭VI	V^7
♭III	♭VII	♭VI - V^7	i

The well-known song 'Amazing Grace' uses inversions, and a secondary dominant, in this case the V of the subdominant (IV). It sounds much like the I chord, only with an added 7th, pointing upward by a fourth. In 'Greensleeves' we have a truly modal harmonic progression, borrowing freely from the parallel major, and briefly pointing to the relative major in the third and fourth phrases. There are many ways to harmonize these songs, these examples present only one possibility.

AROUND THE WORLD
in four songs

Every system of musical notation is essentially a set of instructions for the implementation of sounds through time. In each case they are a kind of time-line, tracking sonic events, and in the case of songs, their marriage to words. Lines, dashes, slashes, curves, numbers, letters, dots, and circles all are used to mirror the up and down inflections, gestures, and shapes of melodies.

Earlier in history, music was an entirely oral tradition, much like storytelling, another narrative art form. As humanity spread and grew, new methods were needed to communicate music to more people. Eventually, notation helped bring music into the homes of everyday musicians, and preserve it for future generations to enjoy. This parallels the development of the printing press, with the same advantages facilitated by that invention. Since we all have basically the same set of musical and linguistic sounds available to us, these various notation systems (*four examples shown opposite*) have a great deal in common. They all indicate the placement of rhythms, the association of notes with syllables of language, and the melodic contours, as well as the overall form of the composition.

In India, because of the complexities of tuning, there are 22 possible tones, allowing for purer acoustic relationships, with simple whole-number ratios between them. Scales are derived from the overall set, depending upon the *raga* desired. Each 7-note *that* or *mela* has a distinct flavour, not unlike the Western modes with its *swaras* (*Sa, Re, Ga, Ma, Pa, Da* and *Ni*) born from the twenty-two *shrutis*.

Bhatkhande notation – Indian
Early 20th century
This is used to indicate the rag, tal,
and tempo, as well as melody and
lyrics for Indian music.

Song of Seikilos – Greek
ca. 200 BC – 100 AD
Likely the oldest surviving music notation
in the world, this example indicates the
lyrics and the basic melodic outline.

Jianpu – Chinese
18th century
This system uses numbers corresponding to pitches
of the scale and dots and lines to indicate durations
and rests, with lyrics streaming underneath.

Medieval Europe
13th century
A musical staff is used here, much like contemporary
staves in Western music, with squares and lines to indicate
pitch and rhythm, and lyrics streaming underneath.

ADVANCED HARMONIES
rascals and spices

Because the dominant chord is the chord of hope and anticipation, chromatic added tones are easily accommodated to add more intervallic complexity. This in turn strengthens the urge to resolve, which can either be fulfilled or denied for the manipulation of tension and release. Extensions can also be added to any of the four chord qualities: major, minor, diminished, or augmented.

It is the non-chord tones or non-station notes that provide the colouring of the essential chord qualities. When they are voiced next to nearby stations, they are considered an 'added' effect. When these colour tones are transposed an octave higher, they become the 9ths, 11ths, and 13ths, which are generally arrived at by stacking thirds. From the root, we pass through the 3rd, 5th, 7th, 9th, 11th, 13th, and conceivably beyond (*opposite top*). These pitch arrangements can yield some startlingly complex harmonic structures, yet the bottom three members of the stack still retain their identity, and imbue the whole edifice with a basic flavour. Music since the end of the 19th century has explored these expanded harmonic possibilities, particularly jazz.

Sixth chords are a category of harmonies with chromatic alterations that don't fit into the usual parallel major/minor borrowing structure, sometimes referred to as vagrants. They are alterations of the subdominant chord, decorating the dominant.

An example of chord notation with extensions is shown below.

MISTY (A) Ebmaj7 | Bbm9   Eb13 | Abmaj7 | Abm9   Db13#11

Ebmaj7 C7#9 | Fm7   Bb7 | Gm7   C7b9 | Fm7   Bb13b9

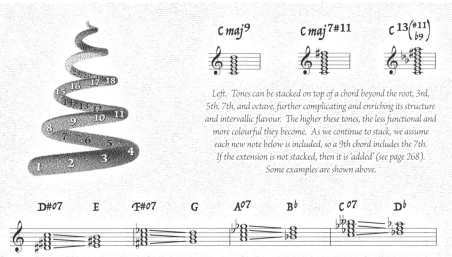

$C maj^9$ $C maj^{7\sharp11}$ $C 13\binom{\sharp11}{\flat9}$

Left. Tones can be stacked on top of a chord beyond the root, 3rd, 5th, 7th, and octave, further complicating and enriching its structure and intervallic flavour. The higher these tones, the less functional and more colourful they become. As we continue to stack, we assume each new note below is included, so a 9th chord includes the 7th. If the extension is not stacked, then it is 'added' (see page 268). Some examples are shown above.

$D\sharp o7$ E $F\sharp o7$ G A^{o7} $B\flat$ C^{o7} $D\flat$

Above: Multiple resolutions of the diminished 7th chord. Because the diminished chord is completely symmetrical, comprised entirely of minor thirds and tritones, any and all of its pitches can function as potential roots. In each example the same four notes in each inversion yield a slightly different enharmonic spelling of the same diminished chord, preserving the syntax of thirds. Each tone has an opportunity to be a leading tone, and resolve upwards by a half step, and can therefore resolve to four possible chords. These resulting four roots themselves in turn spell out a diminished chord, built from consecutive minor thirds (E-G-B♭-D♭).

Italian Aug 6th   German Aug 6th   French Aug 6th   Neapolitan 6th

Above: Resolutions of the various altered 6th chords. The Italian, French, and German versions function like secondary dominants, substituting for the dominant. In the Italian, the augmented sixth is created between the A♭ and the F♯, since they resolve outwards. The German chord is in possession of a perfect fifth, while the unusual flavour of the French is due to the raised fourth, making a chord with two major thirds and two augmented fourths or tritones, a harmonically suggestive symmetrical chord. The Neapolitan version functions like the subdominant, most often a ♭II chord, popular in minor.

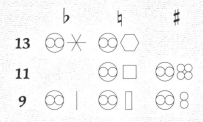

13

11

9

Left: Just as the intermediary tones 2, 4, and 6 come in certain flavours, their correlates, up one octave, the 9th, 11th, and 13th, come in the same flavours, lowered, natural, or raised. All tones above the octave correspond to the tones within the octave, with seven steps added.

ADVANCED FORMS
getting organized

Musical forms are often mixed and matched, with historical hybrids commonplace. Molecular binary and ternary units can be combined and compounded to yield more complex structures, these storyboard squares being used flexibly as general templates.

Attention tends to be highest at the outset of a composition, so music often contains the most intellectually demanding material at this time. An initial tempo will be lively (allegro), perhaps preceded by a slower introduction. Middle movements are often contemplative and reflective, a break from the first movement. Finales are generally light, playful, and dance-like. This common template derives from Baroque dance suites, which were an assemblage of these different moods, variously extroverted and introverted.

There is a general format in the rhetorical unfolding of a complex form: exposition, contrast, development, and summation, and occasionally transformation. Often there is a climax, or a series of them with progressively higher peaks and summits, finally followed by a coming down, unraveling, or denouement.

Sonata form reached its apex in the Classical era and is still in use today. It possesses a fixed relationship of keys and themes. After an introduction, a theme is presented, followed by a contrasting theme. The two themes are then deconstructed and combined in a development section, often tonally unstable or ambiguous, after which the two themes return, a recapitulation. However, the second theme, while first presented in the dominant (or at times another related key) now returns in the tonic key, tying together the journey of contrast and differentiation.

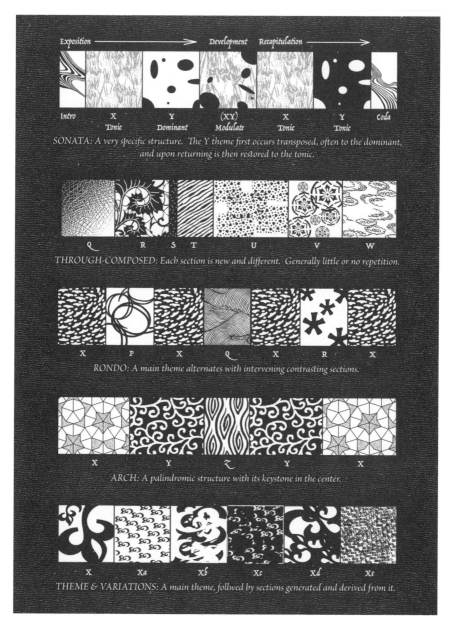

Exposition ⟶   Development   Recapitulation ⟶

Intro   X   Y   (XY)   X   Y   Coda
        Tonic   Dominant Modulate Tonic   Tonic

SONATA: A very specific structure. The Y theme first occurs transposed, often to the dominant, and upon returning is then restored to the tonic.

Q   R   S   T   U   V   W

THROUGH-COMPOSED: Each section is new and different. Generally little or no repetition.

X   P   X   Q   X   R   X

RONDO: A main theme alternates with intervening contrasting sections.

X   Y   Z   Y   X

ARCH: A palindromic structure with its keystone in the center.

X   Xa   Xb   Xc   Xd   Xe

THEME & VARIATIONS: A main theme, follwed by sections generated and derived from it.

287

PUTTING IT ALL TOGETHER
conceive, create, and compose

Perceiving music as an unfolding of epigrams, molecules of meaning, unities of opposites, allows for a new appreciation of its narrative quality. It becomes possible to perceive or even measure the rate of transference as the music unfolds, and to appreciate more deeply the way in which the drama is reinforced or denied, by seeing the very mechanisms by which it accomplishes those things. Degrees of contrast and repetition can be measured, and most importantly variation and transformation can be understood as a kind of evolution. Like the stations of the scale and the pulse, epigrams also have gravity, and what happens in between them communicates part of the drama of the intentionality of the transference; its story, its plight. And in the most skillful hands, our souls follow suit.

In melody these nuances are most easily heard in the large variety of scales found around the world. Although they all possess some forms of fifths and thirds, it is the notes between that convey the real meaning, tension and release, the distances of those 'between' tones, and how they are rhythmically placed. Again, in rhythm, though there is often a predictable pulse, what happens in the spaces in between is much more complex, and can suggest tensions against the different beats, based on their relative distances from the pulse. These epigrams, melodic and rhythmic, can then be arranged into larger coherent structures, compositions, also unifying opposites, and so music is born.

The music we love is the drama of the transference of epigrams, opposites in interplay, unfolding, repeating, contrasting, and most importantly, varying, through melody, harmony, and rhythm.

Chopin Prelude Op. 28, No. 7

Above: A Chopin Prelude. The key is A, the tempo Andantino, a little walking pace, the meter ¾, to be played dolce (sweetly). Curved lines (slurs) indicate the phrases. The melody begins with a pickup on beat three, and commences in the first complete measure. An E⁷ chord suggests the key of A. The C# in measure 1 is an appoggiatura, along with its dotted rhythm followed by three repeated chords, the primary epigram for this little piece. On beat three of measure 2 is a pair of escape tones, followed by a pair of appoggiaturas in measure 3. The harmony proceeds from V to I, completing one full phrase. Measure 5 again discloses the first epigram, with the appoggiatura on beat 1, doubled, now with a dominant 9th chord, elaborating the earlier dominant 7th. The 3rd, G#, is now missing. The second phrase completes the first by measure 8. Measure 9 begins the first phrase again, a literal repetition. Measure 11 sounds much like measure 3, voiced slightly higher in range, but resolves to the surprising secondary dominant F#⁷, the densest chord in the piece. Now it is time to head home. In measure 13 there is a passing tone on beat 1, followed by the falling 7th of B minor, and a voicing of the dominant 9th, this time with the third, G#, included (thank you, Chopin). The last phrase of the last two bars is the final gesture, widely spaced, with the tonic at the top, letting us know the music has indeed finished.

BOOK VI

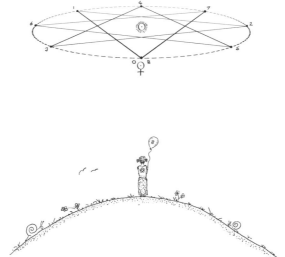

*Climb up a hill at midday on your birthday every year and look at
the Sun. Each year Venus will be three-eighths of the way further
round the Sun, drawing a perfect octagram over eight years.
Now look at the diagram on page 43.*

A Little Book of

COINCIDENCE

IN THE SOLAR SYSTEM

John Martineau

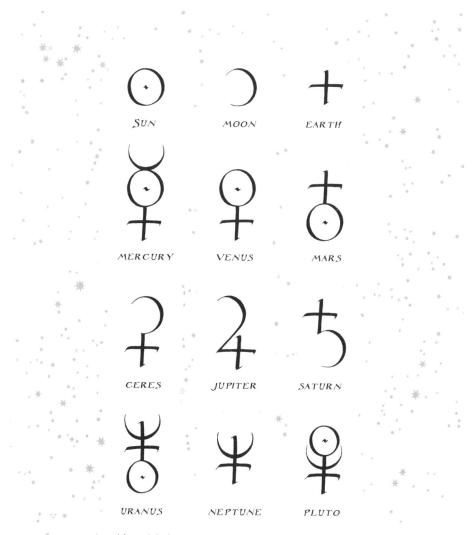

A useful set of glyphs for the planets drawn by calligrapher Mark Mills, each made from Sun, Moon and Earth and used throughout this book.

INTRODUCTION

Biological life is now thought to have appeared on this planet not long after its formation. It seems that the bacterial seeds for the process may have flown in on the tail of a comet or meteor. Speculation is again rife about life under the surface of Mars, on Jupiter's icy moon Europa and indeed anywhere the sacred substance of liquid water is known to exist. The latest plots of the universe depict a structure eerily resembling a vast neural network, suggesting the possibility of a cosmic mind, an ancient idea currently undergoing a renaissance.

The science of the cosmos has changed immeasurably since the Greek and medieval visions of circles of planetary spheres. But despite all recent scientific advances the Earth remains a modern mystery. No convincing theory yet exists to explain the miracle of conscious life, nor the numerous cosmic coincidences which surround our planet. Perhaps the two things are related. This book is not just another pocket guide to our solar system, for it suggests there may be fundamental relationships between space, time, and life which have yet to be understood.

These days we scan the skies listening for intelligent radio signals and looking for other planets similar to our own. Meanwhile, our closest planetary neighbours make the most exquisite patterns around us, in space and in time, and no scientist has yet explained exactly why. Is it really *all just a coincidence*? Why *are* the Sun and the Moon the same size in the sky? How come Venus plays the same numbers around us as plants display on Earth? Read on and see what you think.

GALACTIC DUST
the well-tuned universe

There's a lot going on in the universe. We can now see as many stars within our space-time horizon-bubble as there are grains of sand on Earth. Our planet and we ourselves are made from reorganized smoky stardust, a fact long taught by ancient cultures. We now know that stardust itself is made simply from fizzballs, highly tuned flickering whirlpools of light, long ago squeezed together deep inside stars. We ourselves live in between the little and the large, in a time and a place in the universe where things have condensed, crystallized, built up, tuned in, and settled down.

Just how special are we and our Earth? Funnily enough, scientists are currently puzzling over the strange fact that the whole *universe* seems special. There is *exactly* enough material in the universe to stabilize it, and the ratios between the fundamental forces seem *specifically* organised to produce an amazingly complex, beautiful, and enduring universe. Fiddle with any of the constants, even slightly, and you get a universe of black holes, insubstantial fizzballs, or other lifeless set-ups. Is this design or coincidence? Perhaps our universe is the child of successful parents, who imparted this structure. Maybe the whole quantumly entangled show really is conscious, as Plato taught.

The story of the search for order, pattern, and meaning in the cosmos is very old. The planets of our solar system have long been suspected of hiding secret relationships. In antiquity students of such things pondered the *Music of the Spheres*, today they similarly experiment with the simple precision of Kepler's, Newton's and Einstein's laws.

Who can guess what will come next?

THE SOLAR SYSTEM
spirals everywhere

Our solar system seems to have condensed from the debris of an earlier version some five billion years ago. A Sun ignited in the center and remaining materials were attracted to each other to form small rocky asteroids. Lighter gases were blown out by the solar wind to condense as the four gas giants, Jupiter, Saturn, Neptune, and Uranus while in the inner solar system asteroids grew into planets, the final pieces flying into place with more and more energy as the sizes grew (many still have molten cores today from these collisions). Orbital resonances pushed and pulled planets into new orbits and our solar system eventually took the form of stable disk we see today.

The plane of the solar system is tilted at roughly 30° to the plane of the galaxy and it corkscrews its way around the arm of the milky way. The picture (*opposite top, after Windelius & Tucker*) is schematic of the motions of the four inner planets.

Another way to picture the solar system is by thinking of space-time as a rubber sheet with the Sun as a heavy ball and planetary marbles placed on it (*lower, opposite, after Murchie*). This is Einstein's model of the way matter curves space-time and helps visualize the force of gravity between masses. If we flick a tiny frictionless pea onto our sheet, it could easily be captured by one of the marbles, or be spun around a few times and spat out, or settle into a fast spinning elliptical orbit halfway down any one of the worm-holes. Like a planet, the further the pea gets down the funnel, the faster it must circle to stop itself going down the tube. Also, the faster it spins the heavier it gets and the slightly slower its clocks run.

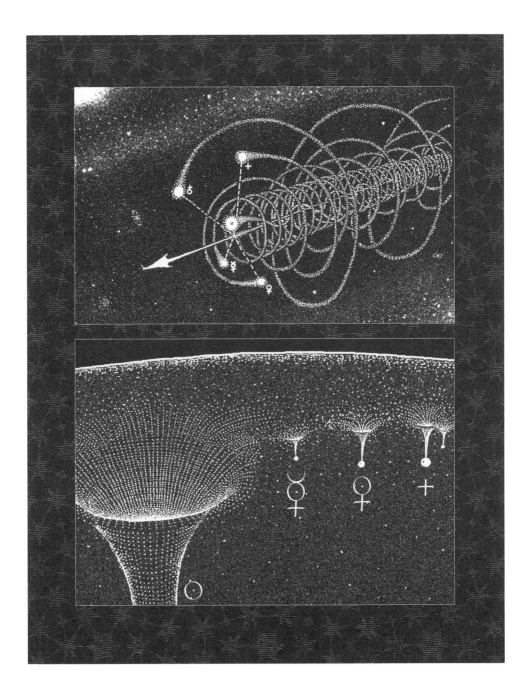

RETROGRADE MOTION
running kissing around

Ancient astronomers who watched the sky from Earth noticed that apart from the Sun and Moon there were five easily visible points of light which moved across the stars. These are the planets, which seem to move around the earth roughly following the Sun's yearly circle, the *ecliptic* or the stars of the *zodiac*. If only life was this simple! Watch planets for any length of time and, far from moving in any simple way, they lurch around like drunken bees, waltzing and whirling. As two planets pass, or kiss, each appears to the other to *retrogress* or go backwards against the stars for a certain length of time.

The diagram below shows Mercury's pattern around a tracked Sun over a year as seen from Earth (*after Schultz*), and opposite we see Cassini's 18th century sketch of the movements of Jupiter and Saturn as seen from Earth. In ancient times hugely complex systems of circles and wheels were called into play to try to mimic these planetary motions (*opposite, below*), culminating in the Ptolemaic system of 39 *deferents* and *epicycles*, used to model the motions of the seven heavenly bodies over two thousand years ago.

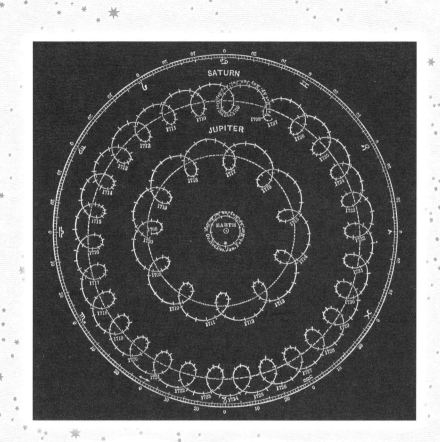

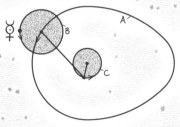

Until 400 years ago planetary motions were modelled using a 'deferent' (A) and an 'epicycle' (B). Other tricks refined the system - here a kind of crank (C) called a 'movable eccentric' produces an egg-shaped deferent for Mercury's dance.

CALENDARS

synchronizing the Sun and Moon

The Sun and Moon may appear perfectly balanced in the sky, but in practice they play a complex pattern which has vexed many cultures over many years. The 29.53 days that occur between full moons are modeled in the Chinese calendar by having alternating months of 29 and 30 days. Similarly, at Stonehenge we find 29 full-width stones and one half-width stone in the sarsen circle to represent 29.5 days.

Many devices hide calendrical themes. For instance a pack of playing cards can be viewed as four seasons, each of $1+2+3+4+5+6+7+8+9+10+11+12+13=91$ days, 364 in all, with the joker as the 365th day. The Tarot likewise conceals secrets, as the Moon and Sun are assigned to the numbers 18 and 19, which, as we shall see on page 331, are indeed the two numbers which best define the calendar.

Perhaps the most extraordinary system ever devised for marrying the heavenly cycles to earthly endeavours is the series of intermeshed calendars developed by the ancient Maya. By 900 AD they had modeled most of the visible solar system with just three cycles: the 365-day Haab, 260-day Tzolkin and enigmatic 819-day cycle. Geoff Stray's incredible diagram opposite is a testimony to just how far scientists are prepared to go to make it all make sense.

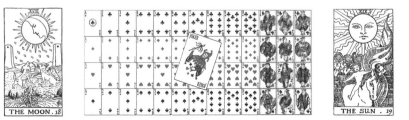

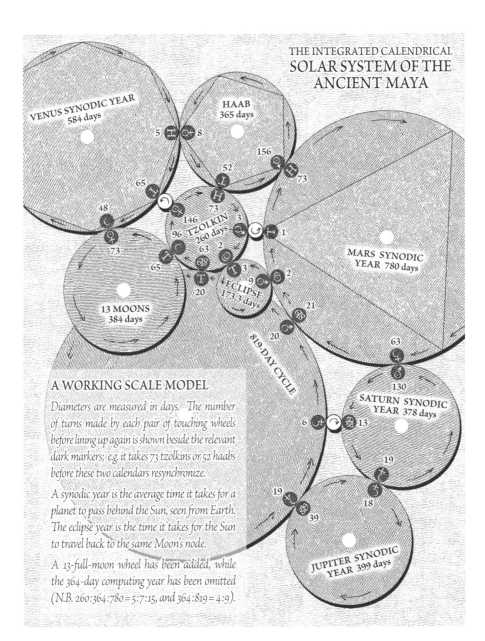

VENUS SYNODIC YEAR
584 days

HAAB
365 days

MARS SYNODIC
YEAR 780 days

TZOLKIN
260 days

ECLIPSE
173.3 days

819-DAY CYCLE

13 MOONS
384 days

SATURN SYNODIC
YEAR 378 days

JUPITER SYNODIC
YEAR 399 days

A WORKING SCALE MODEL

Diameters are measured in days. The number of turns made by each pair of touching wheels before lining up again is shown beside the relevant dark markers; e.g. it takes 73 tzolkins or 52 haabs before these two calendars resynchronize.

A synodic year is the average time it takes for a planet to pass behind the Sun, seen from Earth. The eclipse year is the time it takes for the Sun to travel back to the same Moon's node.

A 13-full-moon wheel has been added, while the 364-day computing year has been omitted (N.B. 260:364:780 = 5:7:15, and 364:819 = 4:9).

The Secret of Sevens
planets, metals and days of the week

A short four hundred years ago the diagrams opposite still formed the cornerstone of cosmological scientific and magical thought across the western world, as they had done for many thousands of years. Today these emblems of the sevenfold system of antiquity appear as quaint reminders of an alchemical cosmology now buried beneath newly discovered planets and physical elements.

There are seven clearly visible wandering heavenly bodies, and they may be arranged around a heptagon in order of their apparent speed against the fixed stars. The Moon appears to move fastest, followed by Mercury, Venus, the Sun, Mars, Jupiter, and Saturn (*top left*). Planets were assigned to days, still clear in many languages, and the order of the days was given by the primary heptagram shown (*top right*). In English, older names for some planets (or gods), were used, thus we have *Wotan's day, Thor's day,* and *Freya's day.*

In antiquity the seven known metals were held to correspond with the seven planets, their compounds giving rise to colour associations. Venus, for example, was associated with the greens and blues of copper carbonates. Students of alchemy would often ponder these relationships as they forged ever more subtle things. Incredibly, the ancient system also gives the *modern* order by atomic number of these metals! Follow a more open heptagram to give *Iron 26, Copper 29, Silver 49, Tin 50, Gold 79, Mercury 80,* and *Lead 82 (lower left after Critchlow & Hinze).* The electrical conductivity sequence also appears round the outside starting with Lead.

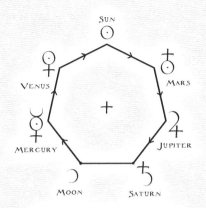

THE SEVEN HEAVENLY BODIES:
Start at the Moon and follow the arrows to give the 'Chaldean Order' of the spheres.

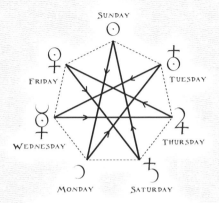
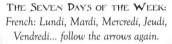

THE SEVEN DAYS OF THE WEEK:
French: Lundi, Mardi, Mercredi, Jeudi, Vendredi... follow the arrows again.

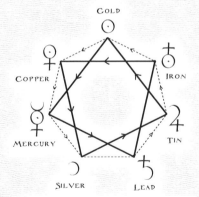

THE SEVEN METALS OF ANTIQUITY:
Start with Iron and follow the arrows to give elements of increasing atomic number.

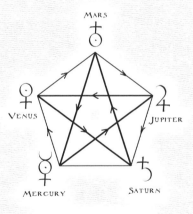

THE FIVE WANDERERS:
Start with Mercury. Moving round the pentagon increases distance from the Sun.

GEOCENTRIC OR HELIOCENTRIC
Earth or Sun at the center

The extraordinary Ptolemaic world of epicycles and deferents lasted a surprisingly long time. Despite its complexity it 'saved appearances' and was also said to save souls. Ellipses were in fact studied by early Greek mathematicians such as Appollonius, and as early as 250 BC Aristarchus of Samos was proposing a system of planets orbiting the Sun. However, it was not to be, and for one and a half thousand years the Earth remained in the center of the universe, just as we experience it, as the Ptolemaic system was handed down from the Greeks to the Arabs, and then back to the West again.

Four early systems are shown opposite (*after Koestler*), and each sphere of each diagram is to be understood as having its own attachment of epicycles and eccentrics. Copernicus, despite in 1543 placing the Sun in the center (*top left*), remained a devout epicycle man, increasing the number of invisible wheels from the Ptolemaic 39 up to an amazing 48. In the late sixteenth century Tycho de Brahe desperately tried to keep the Earth stationary in the center of the universe (*bottom left*), whilst an early Greek model by Herakleides, like a later version by Eriugina, attempted a compromise.

During the 1600s the Sun became the center of the solar system and many people began to forget that planets sometimes move backwards. The modern model of the Solar System (*lower, opposite*) has the planets (including an asteroid, Ceres), orbiting the Sun in ellipses, each planet's ellipse slowly spinning to create a torus or orbital 'shell' over time.

This basic model was first conceived by Johannes Kepler in 1596 and it is to his ideas that we now turn.

Aristarchus & Copernicus

Ptolemy

Tycho de Brahe

Herakleides

Kepler

KEPLER'S VISIONS
ellipses and nested solids

Kepler noticed three things about planetary orbits. Firstly that they are ellipses (*so that a + b = constant, lower, opposite*), with the Sun at one focus. Secondly, that the *area* of space swept out by a planet in a given time is constant. Thirdly, that the period T of a planet relates to R, its semi-major axis (or 'average' orbit), so that T^2/R^3 is a constant throughout the entire solar system.

Looking for a geometric or musical solution to the orbits, Kepler observed that six heliocentric planets meant five intervals. The famous geometric solution he tried was to fit the five *Platonic Solids* between their spheres (*opposite, and detailed below*).

In recent years, far from diminishing Kepler's vision, Einstein's laws actually showed that the tiny space-time effects caused by Mercury's faster (and therefore heavier and time-slowed) motion when nearer to the Sun create a precessional rotation of the ellipses over thousands of years, thus reinforcing Kepler's shells.

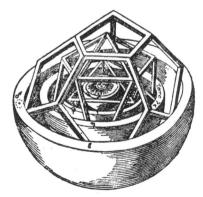

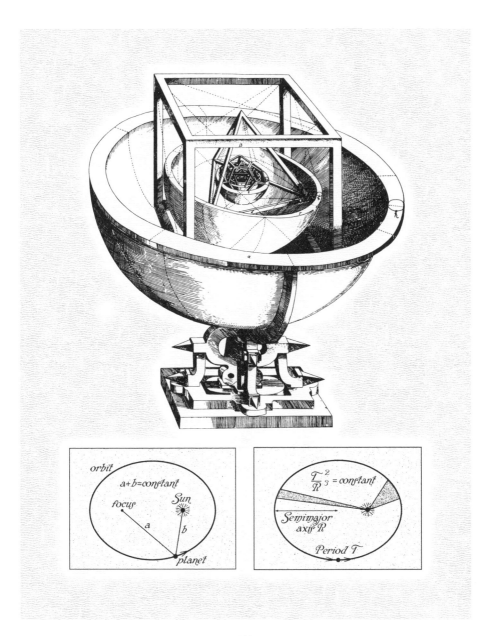

orbit

$a+b=$ constant

focus

Sun

a

b

planet

$\dfrac{T}{R}\dfrac{2}{3}=$ constant

Semimajor axis *R*

Period *T*

THE MUSIC OF THE SPHERES
planets playing in tune

In ancient times the seven musical notes were assigned to the seven heavenly bodies in various symbolic arrangements (*opposite top*). With his accurate data, Kepler now set about precisely calculating these long imagined *Harmoniae Mundi*. He particularly noticed that the ratios between planets' extreme angular velocities were all harmonic intervals (*opposite center, after Godwin*). More recently, work by Molchanov has shown that the entire solar system can be viewed as a 'tuned' quantum structure, with Jupiter as the conductor of the orchestra.

Music and Geometry are close bedfellows and Weizsacker's theory of the condensation of the planets (*opposite after Murchie & Warshall*) throws yet more dappled light on to these elusive orbits. It might appear fanciful were it not for the fact that two nested pentagons (*below left*) define Mercury's shell [99.4%], the empty space between Mercury and Venus [99.2%], Earth and Mars' relative mean orbits [99.7%], and the space between Mars and Ceres [99.8%], while three nested pentagons (*below right*) define the empty space between Venus and Mars [99.6%] and also Ceres and Jupiter's mean orbits [99.6%].

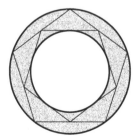 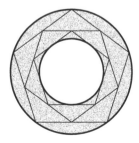

ANCIENT EGYPTIAN SYSTEM

CICERO - SCIPIO'S DREAM

KEPLER'S HEAVENLY HARMONIES

BODE'S LAW AND SYNODS
harmonics and rhythmic kisses

There have been numerous attempts to discover patterns in the orbits and periods of the planets. A basic logarithmic graph (*opposite top*) shows clear underlying order (*after Ovendon & Roy*).

A famous system is the 1750 *Titius & Bode Rule*: To the series 0, 3, 6, 12, 24, 48, 96, 192 & 384, four is added, giving 4, 7, 10, 16, 28, 52, 100, 196 & 388. These numbers fit the planetary orbital radii really quite well (except for Neptune). The formula predicted a missing planet at 28 units between Mars and Jupiter and on 1st January 1801 Piazzi discovered Ceres, the largest of the asteroids in the asteroid belt, in the correct orbit.

The length of time it takes a planet to go once round the Sun is known as its *period*. Sometimes periods occur as simple ratios of each other, a famous example being the 2:5 ratio of Jupiter and Saturn [99.3%]. Uranus, Neptune and tiny Pluto are especially rhythmic and harmonic, displaying a 1:2:3 ratio of periods, Uranus' and Neptune's adding to produce Pluto's [99.8%].

Like a whirlpool, inner planets orbit the Sun much faster than outer planets and the table (*opposite, below*) shows the number of days between two planets' kisses, passes or near approaches, properly called *synods*. Does Earth experience any harmonics? Well, we have two planetary neighbours, Venus sunside and Mars spaceside and the figures reveal that we kiss Mars *three* times for every *four* Venus kisses [99.8%]. So an ultraslow 3 against 4 polyrhythm or a deep musical fourth is being played around us *all* the time!

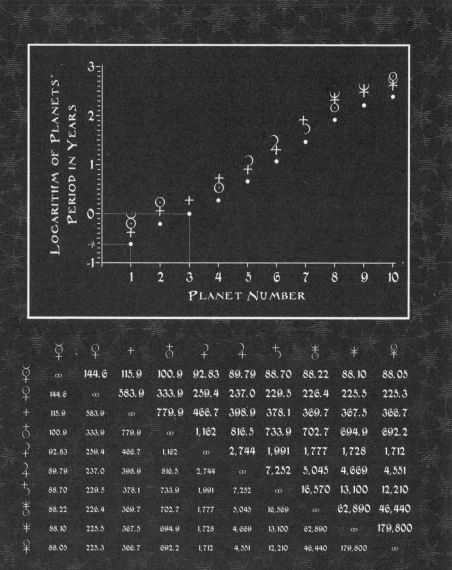

LOGARITHM OF PLANETS' PERIOD IN YEARS

PLANET NUMBER

	☿	♀	+	♁	♃	♃	♄	♅	♆	♇
☿	∞	144.6	115.9	100.9	92.83	89.79	88.70	88.22	88.10	88.05
♀	144.6	∞	583.9	333.9	259.4	237.0	229.5	226.4	225.5	225.3
+	115.9	583.9	∞	779.9	466.7	398.9	378.1	369.7	367.5	366.7
♁	100.9	333.9	779.9	∞	1,162	816.5	733.9	702.7	694.9	692.2
♃	92.83	259.4	466.7	1,162	∞	2,744	1,991	1,777	1,728	1,712
♃	89.79	237.0	398.9	816.5	2,744	∞	7,252	5,045	4,669	4,551
♄	88.70	229.5	378.1	733.9	1,991	7,252	∞	16,570	13,100	12,210
♅	88.22	226.4	369.7	702.7	1,777	5,045	16,569	∞	62,890	46,440
♆	88.10	225.5	367.5	694.9	1,728	4,669	13,100	62,890	∞	179,800
♇	88.05	225.3	366.7	692.2	1,712	4,551	12,210	46,440	179,800	∞

THE INNER PLANETS
Mercury, Venus, Earth and Mars

Our solar system can be thought of a series of thin rotating rings, each slowly settling down. Divided by an asteroid belt into two halves, the inner region sports four small rocky planets quickly orbiting the Sun, while the outer half has four slow huge gas and ice planets.

The Sun has still not given up its secrets. Mostly Hydrogen and Helium, and an element factory, it is also a giant fluid geometric magnet, 15 million°C at its core, 6,000°C at the surface. It blows a particle wind through the entire solar system and its sunspots and huge solar flares affect electronics on Earth.

Mercury is the first planet. Mostly solid iron, it is a cratered, atmosphereless world, 400°C in the sunshine, -170°C in the shade.

Venus is second, a cloud-shrouded greenhouse world. On the surface the temperature is a staggering 480°C and the carbon-dioxide rich atmosphere is *ninety* times denser than Earth's. An apple here would be instantly incinerated by the heat, crushed by the atmosphere and finally dissolved in sulphuric acid rain.

Earth is the third planet, the one with life and one moon.

Mars is fourth, a rocky red world, just above freezing. Ice caps cover the poles under a thin atmosphere. River beds suggest that Mars may once have had oceans but they are long gone now, and today dust storms regularly envelop the planet for days. Huge dead volcanoes, one three times larger than Mount Everest, stand witness to a bygone age. Mars has two tiny moons.

Beyond Mars is the Asteroid Belt, and, beyond that, the giants.

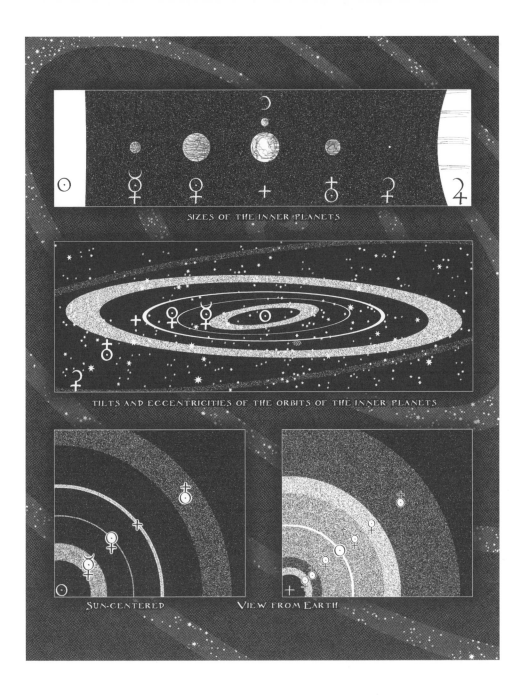

SIZES OF THE INNER PLANETS

TILTS AND ECCENTRICITIES OF THE ORBITS OF THE INNER PLANETS

SUN-CENTERED

VIEW FROM EARTH

MAKING SENSE OF THE PICTURES
a few tips on appearances

Seen from Earth, day or night, the Sun moves slowly to the left against the stars (right in the southern hemisphere), taking a year to return to the same star. The Moon swiftly circles around in the same direction every month, taking 27.3 days to return to a star. Venus and Mercury oscillate around the Sun, coming and going, as the Sun itself slowly trundles around its yearly circle. Imagine standing on Venus—the Sun moves faster against the stars and Mercury is closer, whirling round the Sun like a fairground waltzer.

Every *pair* of planets creates a *single* dance. It doesn't matter which of the two you stand on, your partner's dance around you will be the same. It is a shared experience. Mercury's evolving waltzes with Earth and Venus are shown (*opposite top*). Earth and Mercury roughly kiss 22 times in 7 years, though the ancient Greeks also knew of a more accurate 46 year, 145 synod cycle. Mercury and Venus are beautifully in tune after just 14 kisses.

Shown lower, opposite is the Golden Section, φ or *phi*. It is found throughout every pentagram and in the Fibonacci series of numbers (*opposite*), which starts with 1, 2, 3, 5, 8 and 13, all numbers we will see in the inner planets. The Golden Section is essentially 0.618, but since one *divided* by it is 1.618 (which is the same as *adding* one to it), and 1.618 *times* 1.618 equals 2.618 (the same as adding one more), it often takes any of these values. The Golden Section is found throughout organic life forms; it is the signature of life, and, as we shall see, highly accurately present in the inner solar system too.

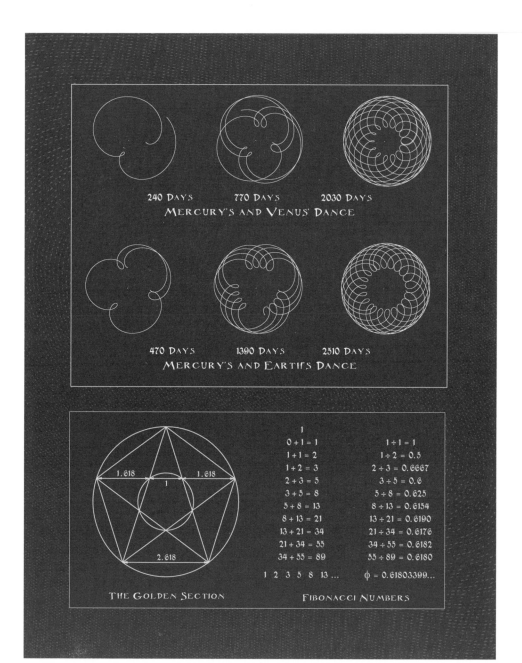

240 DAYS   770 DAYS   2030 DAYS

MERCURY'S AND VENUS' DANCE

470 DAYS   1390 DAYS   2510 DAYS

MERCURY'S AND EARTH'S DANCE

1.618   1   1.618

2.618

$$1$$
$$0 + 1 = 1$$
$$1 + 1 = 2$$
$$1 + 2 = 3$$
$$2 + 3 = 5$$
$$3 + 5 = 8$$
$$5 + 8 = 13$$
$$8 + 13 = 21$$
$$13 + 21 = 34$$
$$21 + 34 = 55$$
$$34 + 55 = 89$$

$$1 \quad 2 \quad 3 \quad 5 \quad 8 \quad 13 \ldots$$

$$1 \div 1 = 1$$
$$1 \div 2 = 0.5$$
$$2 \div 3 = 0.6667$$
$$3 \div 5 = 0.6$$
$$5 \div 8 = 0.625$$
$$8 \div 13 = 0.6154$$
$$13 \div 21 = 0.6190$$
$$21 \div 34 = 0.6176$$
$$34 \div 55 = 0.6182$$
$$55 \div 89 = 0.6180$$

$$\phi = 0.61803399\ldots$$

THE GOLDEN SECTION   FIBONACCI NUMBERS

MERCURY AND VENUS' ORBITS
a very simple aide-memoire

There are few things more simple than a circle. With Kepler's discovery of the ellipses, and Newton and Einstein setting them spinning, the planetary orbits can be thought of as orbital 'circles', centered on the Sun, with the eccentricity thickening the circle slightly, or giving the spheres a shell (*see Kepler's diagram, page 307*).

One of the very first things you can do with circles is to put three of them together so that they all touch. Amazingly, the orbits of the first two planets of the solar system are hiding in this simple design. If Mercury's mean orbit passes through the centers of the three circles then Venus' encloses the figure [99.9%].

This is a simple trick to remember—you see it all around you all the time, in the home, in design, art, architecture and nature. Every time you pick up three glasses or push three balls together you create the first two planets' circular orbits, to an extraordinary degree of accuracy. There must be a reason for this beautiful fit between the ideal and the manifest, but none is yet known and these kinds of problems are currently out of fashion; perhaps a bright 21st century scientist will find an answer—until then it remains a 'coincidence'.

The triangle is one emblem of the musical octave 2:1, and Mercury performs a delightful solo on this theme, as one Mercury day is exactly two Mercury years, during which time the planet has spun on its own axis exactly three times. Thus the very first planet plays the very first harmonies and draws one of the very first geometrical shapes. We have started with one, heard a two, and seen a three.

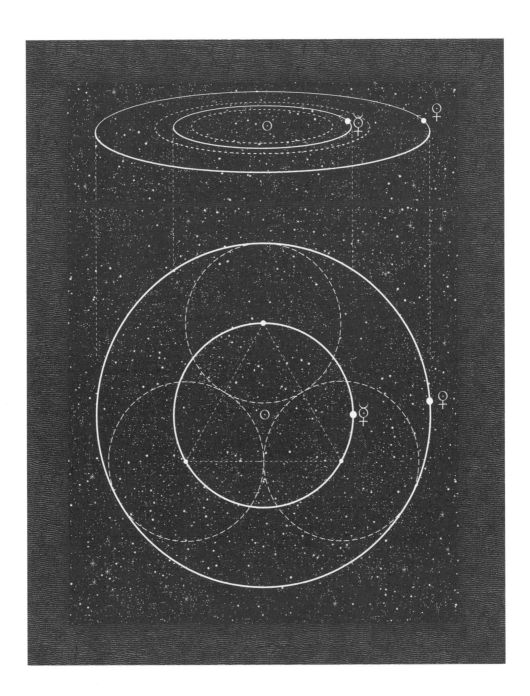

THE KISS OF VENUS
our most beautiful relationship

Other than the Sun and Moon, the brightest point in the sky is Venus, morning and evening star. She is our closest neighbour, kissing us every 584 days as she passes between us and the Sun. Each time one of these kisses occurs the Sun, Venus and the Earth line up two-fifths of a circle further around—so a pentagram of conjunctions is drawn, taking exactly eight years [99.9%], or thirteen Venusian-years [99.9%]. Notice the Fibonacci numbers again, 5, 8, and 13, which govern most plant growth on Earth. The periods of Earth and Venus are also closely related as $\phi:1$ [99.6%].

Seen from Earth this harmony appears as Venus whirling around the trundling Sun drawing an astonishingly beautiful pattern. In the diagram (*opposite top*) four eight-year cycles are shown, so 32 years. The small loops are made when Venus in her closest dazzling kiss seems briefly to reverse direction against the background Stars (*shown below, as seen from Earth*).

The fivefold nature of Venus and Earth's dance extends to their closest and furthest distances from each other, as Venus' *perigee* and *apogee* are defined by two pentagrams (*lower, opposite*). The body of space one draws around the other is thus sized $1:\phi^4$ [99.98%].

All these diagrams *also* apply to Venus' experience of Earth.

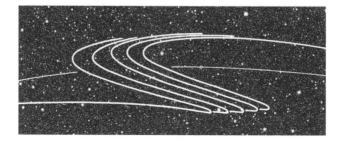

THE PERFECT BEAUTY OF VENUS
the things they don't teach you at school

With the Sun in the center, let us look at the orbits of Venus and the Earth. Every couple of days a line is drawn between the two planets (*below left*). Because Venus orbits faster she completes a whole circuit in the same time that the Earth completes just over a half-circuit (*below center*). If we keep watching for exactly eight years (or 13 Venusian years) the pattern opposite emerges, the sun-centered version of the five-petaled flower on the previous page.

The ratio between Earth's outer orbit and Venus' inner orbit, i.e. their *home*, is intriguingly given by a square (*below right*) [99.9%].

Venus rotates extremely slowly on her own axis in the opposite direction to most rotations in the solar system. Her rotation period is precisely two-thirds of an Earth year, a musical fifth. This closely harmonizes with the dance opposite so that every time Venus and Earth kiss, Venus does so with her *same face* pointing at the Earth. Paint a spot on Venus' surface as she passes in front of the Sun, and every time she lines up with the Sun again as seen from Earth the spot will be pointing at you again. Over the eight Earth years of the five kisses, Venus spins on her own axis twelve times in thirteen of her years (*from Kollerstrom*). All beautiful musical numbers.

PHYLLOTAXIS
the spiral of life

Life on Earth uses one set of numbers above all others. Phyllotaxis is the study of the way leaves are arranged along a stem, and also describes other features of plants such as flowers, seed heads, and fruits. The key to phyllotaxis is the Fibonacci sequence 1, 1, 2, 3, 5, 8, 13, 21, 34, 55 and so on, where adjacent terms define the Golden Section increasingly well, present in pentagrams, which we have met before.

It is a simple fact that most plants on Earth produce alternate leaves at Fibonacci fractions of a full rotation. For example, some plants produce leaves along a stem every ½ rotation, in hazel and beech trees the angle is ⅓, in apricots and oak trees it is ⅖, in pear and poplar trees it is ⅜, and in almond and willow trees it is 5/13. Pineapples display 5-, 8- and 13-armed spirals. Count the number of buds along a sprig of pussy willow and you will find a spiral of 13 buds in 5 turns.

Humans use the same numbers (in a fourfold manner). We have 5 fingers/toes in each quarter, a pattern repeated in our mouths as 5 milk teeth in each quarter, replaced by 8 adult teeth, 13 in all per quarter.

The most common number of petals in flowers is 5, and the most commonly used phyllotaxis numbers in plants are 5, 8, and 13.

As above, so below—for these are also the numbers of Venus!

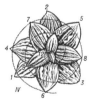

Above left: 8:13 phyllotaxis. New unit elements appear along an Archimedean spiral every 137.5°.
Above right: W Hofmeister's space-jostling model of phyllotaxis based on Leonardo Da Vinci's suggestion that phyllotaxis optimizes access to dew and sunlight.
Left: Pineapples display lovely 5:8:13 phyllotaxis as 5 near-horizontal spirals, 8 45° spirals and 13 verticals.
Below: Many seed heads display phyllotaxis as a result of the phyllotaxis implicit in the petals of their flowers.

MERCURY AND EARTH
yet more phives and eights

Mercury and Earth's physical sizes are in the same relation as their mean orbits! Various five and eightfold overlays are shown opposite which proportion the orbits *and* sizes of these two planets.

The diameter of Mercury's *innermost* orbit is suggested by the pentagram incircle (*top left*) [99.5%] and also happens to be the distance between the mean orbits of the two planets [99.7%].

The diagram bottom right expands on the three touching circles of page 317. Eight circles centered on Venus' orbit produce Earth's mean orbit [99.99%]—the eight years of the five kisses perhaps?

Mercury, Venus, and Earth display a peculiar coincidence: If we work in units of Mercury's orbital radius and period, then Venus' period times 2.618 is Earth's orbital radius squared [99.8%]. Mercury's dance around Earth also produces its synodic year of 115.9 days. Richard Heath recently discovered that this is 2.618 times a musical fifth times a full moon [99.9%]—a musical fifth is 3:2; 2.618 is Φ^2 (or 1.618 × 1.618), and there is a full moon every 29.53 days.

Earth's and Saturn's relative orbits *and* sizes are given by a fifteen-pointed star (*below*), which also produces the tilt of the Earth.

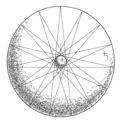

THE RELATIVE SIZES OF MERCURY AND EARTH DEFINED
BOTH BY A PENTAGRAM OR AN OCTAGRAM (99%)

THE RELATIVE ORBITS OF MERCURY AND EARTH DEFINED
BY THE SAME PENTAGRAM AND OCTAGRAM (99%)

OTHER MORE ACCURATE WAYS OF DRAWING THE INNER ORBITS (99.9%)

THE ALCHEMICAL WEDDING

three to eleven all round

From the surface of the Earth, the Sun and the Moon appear *the same size*. According to modern muggle cosmology this is 'just' a coincidence, but any good wizard will tell you the balance between these two primary bodies is clear proof of very ancient magic.

The size of the Moon compared to the Earth is 3 to 11 [99.9 %]. What this means is that if you draw down the Moon to the Earth, then the circle through the center of the heavenly Moon will have a circumference equal to the perimeter of a square enclosing the Earth. As we saw on page 78, this proportion is also present in every double rainbow you see. The ancients seem to have known about this, and hidden it in the definition of the mile (*opposite, after Michell & Ward*).

The Earth-Moon proportion is also precisely invoked by our two neighbours, Venus and Mars (*Venus shown dancing round Mars below*). The closest:farthest distance ratio that each experiences of the other is, incredibly, 3:11 [99.9%]. The Earth and the Moon sit in between them, perfectly echoing this beautiful local spacial ratio.

3:11 happens to be 27.3% and the Moon orbits the Earth every 27.3 days, the same period as the average rotation period of a sunspot.

The Sun and Moon do seem very much the unified couple.

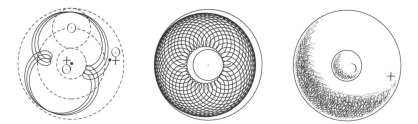

THE MOON, A TOTAL SOLAR ECLIPSE AND THE SUN, AS SEEN FROM EARTH

THE SIZES OF THE MOON AND THE EARTH 'SQUARE THE CIRCLE'
THE DASHED SQUARE AND CIRCLE ARE THE SAME LENGTH OF STRING

MILES OF MOON AND EARTH

RADIUS OF MOON = 1080 MILES = 3 × 360 MILES
RADIUS OF EARTH = 3960 MILES = 11 × 360 MILES
DIAMETER OF MOON = 2160 MILES = 3 × 1 × 2 × 3 × 4 × 5 × 6 MILES
RADIUS OF EARTH + RADIUS OF MOON = 5040 MILES
= 1 × 2 × 3 × 4 × 5 × 6 × 7 = 7 × 8 × 9 × 10 MILES
DIAMETER OF EARTH = 7920 MILES = 8 × 9 × 10 × 11 MILES
THERE ARE 5280 FEET IN A MILE
= (10 × 11 × 12 × 13) − (9 × 10 × 11 × 12)

CALENDAR MAGIC

just three numbers do the trick

Recent work by Robin Heath has revealed simple geometrical and mathematical tools which suggest order and form within the Sun-Moon-Earth system. Imagine we want to discover the number of full moons in a year (somewhere between 12 and 13). Draw a circle, diameter thirteen with a pentagram inside. Its arms will then measure 12.364, the number of full moons in a year [99.95%].

An even more accurate way of doing it is to draw the second Pythagorean triangle, which just happens to be made of 5, 12, and 13 again, the numbers of the keyboard, and of Venus (*page 320*). Dividing the 5 side into its harmonic 2:3 gives a new length, the square root of 153, 12.369, the number of full moons in a year [99.999%].

The Moon seems to beckon us to look further. We all know that six circles fit around one on a flat surface (6 and 7). Twelve spheres pack perfectly around one in our familiar three-dimensional space (12 and 13 again). We seem to be moving up in sixes. Could *eighteen* time-spheres fit around one in a fourth dimension of time? Incredibly, all of the current major time cycles of the Sun-Moon-Earth system turn out to be accurately defined as simple combinations of the numbers 18, 19, and the Golden Section.

The Golden Section is evident in the pentagram, the icosahedron, the dodecahedron and all living things. The orbits of the four inner planets all display its presence. Its values 0.618, 1, 1.618 and 2.618 added to the magic number 18 produce 18, 18.618, 19, 19.618, and 20.618, which then multiply together as shown opposite.

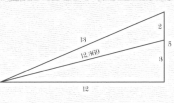

TWO ANCIENT TECHNIQUES
FOR FINDING THE NUMBER
OF FULL MOONS IN A YEAR

18 YEARS = THE SAROS ECLIPSE CYCLE (99.83%)
(SIMILAR ECLIPSES WILL OCCUR AFTER 18 YEARS)

18.618 YEARS = REVOLUTION OF THE MOON'S NODES (99.99%)
(THE MOON'S NODES ARE THE TWO PLACES WHERE THE SLIGHTLY
OFFSET CIRCLES OF THE SUN AND MOON'S ORBITS CROSS)

19 YEARS = THE METONIC CYCLE (99.99%)
(IF THERE IS A FULL MOON ON YOUR BIRTHDAY THIS YEAR - THERE
WILL BE ANOTHER ONE ON YOUR BIRTHDAY IN 19 YEARS TIME)

THE ECLIPSE YEAR = 18.618 × 18.618 DAYS (99.99%)
(THE ECLIPSE YEAR IS THE TIME IT TAKES FOR THE SUN TO RETURN
TO THE SAME ONE OF THE MOON'S NODES. IT IS 18.618 DAYS SHORT
OF A SOLAR YEAR (99.99%). THERE ARE 19 ECLIPSE YEARS IN A SAROS)

12 FULL MOONS = 18.618 × 19 DAYS (99.82%)
(12 FULL MOONS IS THE LUNAR OR ISLAMIC YEAR)

THE SOLAR YEAR = 18.618 × 19.618 DAYS (99.99%)
(THE SOLAR YEAR IS THE 365.242 DAY YEAR WE ARE USED TO)

13 FULL MOONS = 18.618 × 20.618 DAYS (99.99%)
(13 FULL MOONS IS ANOTHER 18.618 DAYS AFTER THE SOLAR YEAR)

COSMIC FOOTBALL

Mars, Earth and Venus spaced

The next planet out from Earth is the fourth planet, Mars. Kepler had tried a *dodecahedron* spacing the orbits of Mars and Earth and an *icosahedron* spacing Earth from Venus (*see page 306*), and, coincidentally, it turns out he was very close to the mark.

The dodecahedron (made of twelve pentagons) and the icosahedron (made of of twenty equilateral triangles) are the last two of the five Platonic Solids (*see BOOK III of this volume*). They form a pair, as each creates the other from the centers of its faces (*below*). Opposite, they appear in bubble form inside Mars' spherical mean orbit. The dodecahedron magically produces Venus' orbit as the bubble within (*opposite top*) [99.98%], while the icosahedron defines Earth's orbit through its bubble centers (*lower, opposite*) [99.9%].

In the ancient sciences the icosahedron was associated with the element of Water, so it is appropriate to see it emanating from our watery planet. The dodecahedron represented *aether*, the life force, here enveloping lively Earth, and defined by her two neighbours.

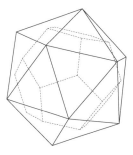

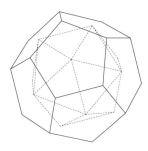

VENUS AND
MARS

EARTH
AND MARS

THE ASTEROID BELT
through the looking glass

We have reached the end of the inner solar system. Beyond Mars lies a particularly huge space, the other side of which is the enormous planet Jupiter. It is in this space that the Asteroid Belt is found, thousands of large and small tumbling rocks, silicaceous, metallic, carbonaceous, and others. There are spaces, *Kirkwood Gaps*, in the asteroid belt, cleared where orbital resonances with Jupiter occur. The largest gap is at the orbital distance which would correspond to a period of one third that of Jupiter.

The largest of the asteroids by a very long way is Ceres, comprising over one third of the total mass of all of them. She is about the size of the British Isles and produces a *perfect* eighteenfold pattern with Earth (*see page 403*).

Bode's Law predicted something at the distance of the asteroid belt (*see page 310*), but it was Alex Geddes who recently discovered the weird mathematical relationship between the four small inner planets and the four outer gas giants. Their orbital radii magically 'reflect' about the asteroid belt and multiply as shown below and opposite to produce two enigmatic constants.

$$Ve \times Ur = 1.204 \ Me \times Ne \qquad Ve \times Ma = 2.872 \ Me \times Ea$$
$$Me \times Ne = 1.208 \ Ea \times Sa \qquad Sa \times Ne = 2.876 \ Ju \times Ur$$
$$Ea \times Sa = 1.206 \ Ma \times Ju \qquad (Ve \times Ma \times Ju \times Ur = Me \times Ea \times Sa \times Ne)$$

The asteroid belt is unlikely to be the remains of a small planet as no sizeable body could ever have formed so close to Jupiter.

THE ASTEROID BELT SEPARATES THE FOUR SMALL ROCKY
INNER PLANETS FROM THE FOUR HUGE OUTER ONES

THE PATTERN OF GEDDES' MAGICAL MULTIPLICATIONS

THE OUTER PLANETS
Jupiter, Saturn, Uranus, Neptune, and beyond

Beyond the Asteroid Belt we come to the realm of the gas and ice giants, Jupiter, Saturn, Uranus, and Neptune.

Jupiter is the largest planet, and its magnetic field is the largest object in the solar system. Ninety percent hydrogen, it is nevertheless built around a rocky core like all the giant planets. Metallic hydrogen and then liquid hydrogen surrounds this core. The famous Red Spot is a storm, larger than Earth, which has raged now for hundreds of years. Jupiter's moons are numerous and fascinating: One, Io, is the most volcanic body in the solar system; another, Europa, may have warm oceans of water beneath its icy surface.

The next planet is Saturn, with its beautiful system of rings. Saturn's structure beneath its clouds is much the same hydrogen and helium mix as Jupiter. A large number of moons have been discovered, the largest of which is Titan, a world the size of Mercury with all the building blocks for life.

Beyond Saturn is Uranus, which orbits on its side. Winds gust on the equator at six thousand times the speed of sound.

Next is Neptune, like Uranus an ice world of water, ammonia and methane. The largest moon, Triton, has nitrogen ice caps and geysers which spew liquid nitrogen high into the atmosphere.

Finally, tiny Pluto, and, beyond that, the primordial swarm of the Kuiper Belt. Then, stretching a third of the way to the nearest star, the sphere of icy debris and comets of the Oort Cloud.

SIZES OF THE OUTER PLANETS

TILTS AND ECCENTRICITIES OF THE ORBITS OF THE OUTER PLANETS

SUN-CENTERED

VIEW FROM THE EARTH

FOURS

Mars, Jupiter and massive moons

An asteroid belt and 550 million km separate Mars' and Jupiter's orbits, further than Earth is from the Sun. Jupiter is the first and largest of the gas giants, the vacuum cleaner of the solar system. If Jupiter had gathered only slightly more material during its long and ongoing formation its internal pressures would have turned it into a star and we would have had a second Sun.

The top diagram opposite shows a simple way to draw the orbits of Mars and Jupiter from four touching circles or a square [99.98%]. It is a proportion commonly seen in church windows and railway stations. Shown below, on this page, is a pattern from the same family, which accurately spaces Earth's and Mars' orbits [99.9%].

Jupiter has four particularly large moons. The two largest, Ganymede and Callisto, are the size of the planet Mercury and produce one of the most perfect space-time patterns in the solar system. An observer living on either moon would experience the motions of the other in space and time as the beautiful fourfold diagram shown opposite.

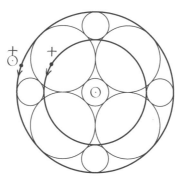

HOW TO ACCURATELY DRAW MARS & JUPITER'S MEAN ORBITS

THE BEAUTIFUL DANCE OF GANYMEDE & CALLISTO

OUTER MOONS
harmonic patterns

Four groups of moons orbit Jupiter. The first two groups have four moons each and look very like a model of the whole solar system—four small inner bodies followed by four big outer bodies. The second group, of four particularly large moons, the *Galileans*, is further divided into two small rocky worlds, Io and Europa, then two gas and ice moons the size of planets, Ganymede and Callisto.

The grouping into fours is striking. Each of the four groups has its own general moonsize, orbital plane, period and distance from Jupiter (the inclinations of the four orbital planes of the four groups even add up to 90°, a quarter of a circle [99.9%]).

Saturn has over thirty moons, most shepherding and tuning the amazing rings with the larger bodies tending to be further out. Far beyond Saturn's rings, however, are three moons—the gigantic Titan, tiny Hyperion and, further out still, Iapetus.

Opposite are shown some harmonic patterns: two from Jupiter's largest moons, two experienced by Saturn's giant moon Titan, and two experienced by Neptune, the outer planet of the solar system.

Dischords are rare. The solar system seems to enjoy harmony.

340

JUPITER'S GIANT SEAL
huge hexagrams and affirmatory asteroids

Jupiter, the largest planet, was king of the ancient gods, Zeus to the Greeks. A delightful feature of its orbit is its pair of asteroid clusters. *The Trojans* are two groups of asteroids which move round Jupiter's orbit, 60° ahead of it and 60° behind (*opposite*). This partnership perpetually moves round the Sun as though held in place by the spokes of a wheel. The positions of the Trojan clusters are known as the *Laplace Points*, with Sun, Jupiter, and Trojans forming gravitationally balanced equilateral triangles.

Just for the fun of it, if we now join the spokes as shown opposite then three hexagrams can be seen to produce Earth's mean orbit from Jupiter's [99.8%]—a very easy trick to remember. Earth and Jupiter's orbits are thus lurking in every crystal. Another name for a six-pointed star made of two triangles is a *Star of David* or *Seal of Solomon*.

Exactly the same Earth-Jupiter proportion may be created by spherically nesting three cubes, or three octahedra, or any threefold combination of them (*e.g., below*). If the outer sphere is Jupiter's mean orbit then the inner one is the sphere of Earth's mean orbit. By Jove!

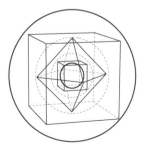 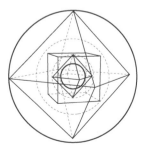

342

THE GOLDEN CLOCK

Jupiter and Saturn seen from Earth

Jupiter and Saturn are the two largest planets of the solar system and ruled the outer two spheres of the ancient system. In ancient mythology, Saturn was Chronos, the Lord of Time.

The top two diagrams opposite show the close 5:2 ratio of their periods. Top left we see their dance; the beautiful threefold harmonic is immediately apparent, spinning slowly because of the slight miss in the harmony. From Earth, this pattern is seen as an important sequence of conjunctions and oppositions of Jupiter and Saturn, who kiss every 20 years. Top right we see the hexagram created by these positions—with conjunctions marked on the outside of the zodiac and oppositions marked inside. The planets move anticlockwise around the dashed circle of the ecliptic, starting at twelve o'clock, Jupiter moving faster than Saturn.

The lower diagram shows the relative speeds of orbit of Earth, Jupiter and Saturn. We start with the three planets in a synodic line at twelve o'clock. Earth orbits much faster than the outer planets and makes a complete circuit of the Sun (365.242 days) and then a bit more before lining up with slowcoach Saturn again for a synod after 378.1 days. Three weeks later it lines up again with Jupiter (after 398.9 days). Richard Heath recently discovered that the Golden Section is defined here in time and space to a *stunning* 99.99% accuracy! The two giants of our solar system thus focus the Golden Section on us, in space and time, reinforcing the geometry of life on Earth.

Less importantly, Saturn takes the same number of years to go round the Sun as there are days between full Moons [99.8%].

JUPITER & SATURN'S DANCE

CONJUNCTIONS & OPPOSITIONS

φ=0.618034

JUPITER & SATURN'S SYNODS DEFINE THE GOLDEN SECTION

OCTAVES OUT THERE
threes and eights again

If you ever want to incorporate Jupiter, Saturn and Uranus' orbits into a window or floor design the diagram opposite might help. An equilateral triangle and an octagram proportion the outer, mean and inner orbits of the three largest planets. Tiny inaccuracies are visible but the fit is excellent overall, memorable, and adequate for many practical purposes. It is a spiky inversion of the touching circles solution for the first three planets (*see page 325, lower right*).

One way of depicting the musical octave (a halving or doubling of frequency or wavelength) is by an equilateral triangle, as the inscribed circle has a diameter half that of the containing circle.

Another rule of thumb is to remember that if Jupiter's orbit is 6, then Saturn's is 11 [99.9%], twice the Moon:Earth size ratio (*page 326*).

Saturn's orbit also happens to invoke π or '*pi*'—twice (*below*): Its radius is the circumference of Mars' orbit [99.9%] and its circumference is the diameter of Neptune's orbit [99.9%].

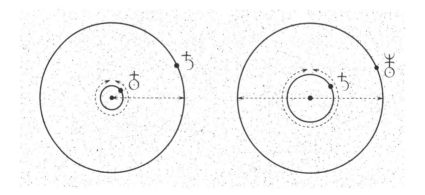

GALACTIC GEOMETRY
to the stars and beyond

Moving further out into the solar system triangular geometries prevail (*two shown below* [99.9%]). Uranus and Neptune, like Saturn, both have mysterious ring systems with clear spaces at Kirkwood distances where particles orbit at periods harmonic with one or more moons. Uranus' bright outer ring has a diameter twice that of Uranus itself [99.9%], echoing the orbits of Uranus and Saturn, and Neptune's innermost ring is two-thirds the size of its outermost [99.9%]. These proportions beautifully invoke the local timing, as Neptune's orbital period is twice that of Uranus, and Uranus' is two-thirds that of Pluto, an outer reflection of the inner harmonic 1:2:3 we saw with Mercury.

One of the most obvious symmetries of modern cosmology occurs in that the Milky Way, i.e. the plane of our own galaxy, is tilted at almost exactly 60° to the ecliptic or plane of our solar system (*shown horizontal opposite*) [99.7%]. What is more, every year the Sun crosses the galaxy through the galactic center, and being alive in these times means this happens on midwinter's day. Like many of the images in in these pages you may need to study this to get it!

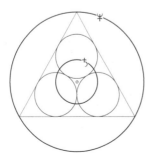

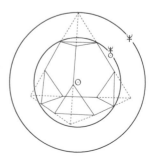

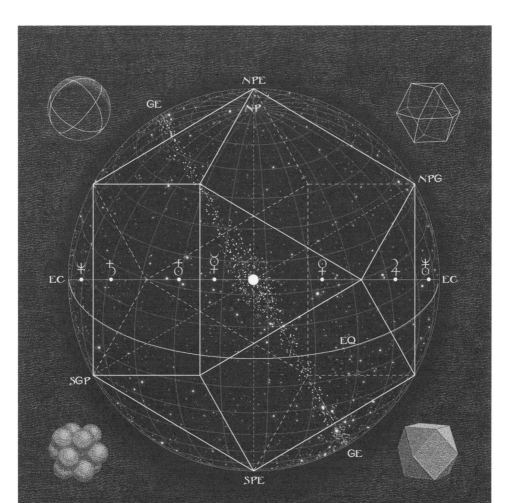

NPE - North Pole of the Ecliptic
GE - Galactic Equator
NP - North Pole of the Earth
NPG - North Pole of our Galaxy
EC - Ecliptic, Path of Sun
EQ - Earth's Equator
SGP - South Pole of our Galaxy
SPE - South Pole of the Ecliptic

It is Winter Solstice and the North Pole of the Earth is tilted away from the Sun which is right in front of the center of our Galaxy. The poles of the Ecliptic and the Galaxy define four points of a Hexagon in space around us.

ICE HALOS
rainbows where planets lie

On certain still afternoons, if you are lucky, you will see a pair of rainbow spots left and right of the Sun. Known as 'Sun dogs', these are the first elements to appear of an ice halo—a thin rainbow circle around the Sun. Caused by light passing through ice crystals high in the atmosphere, Sun dogs appear 22.5° left and right of the Sun, just outside the bright 22° halo. Sometimes, a second larger halo appears 46° from the Sun, with a distinctive arc on top, the whole arrangement looking strangely similar to the ancient glyph for Mercury.

Amazingly, these two ice halos match the mean orbits of the inner two planets Mercury and Venus as seen from the surface of the Earth. This means that when you look at a double ice halo, you really are seeing the spheres of the mean orbits of Mercury and Venus, hanging in the sky. And what is more, the same two ice halos *also* function as a diagram of the relative orbits of Venus and Mars.

This is extraordinary. Every circle fits. Sunlight and ice dust paint orbits as rainbows while the Sun and Moon appear the same size in the sky. Our closest neighbour dances 5-fold around us in 8 years or 13 of her years, while below on Earth plants also dance 5, 8 and 13. These coincidences are focused on us, here, and now, a planet of conscious observers. Perhaps consciousness has played some part in creating this. Could the act of observation lense reality in some way? Do beautiful improbable coincidences surround other planets of observers?

Plato writes that things are more perfectly organised than we can ever imagine. How *do* you balance a Sun and a Moon? Could we in fact be living in a conscious quantum holographic universe?

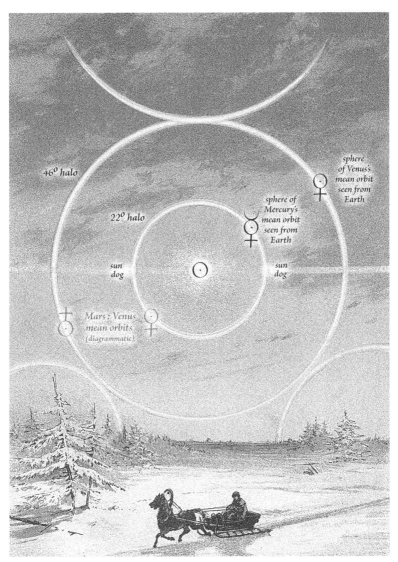

46° halo

22° halo

sphere of Venus's mean orbit seen from Earth

sphere of Mercury's mean orbit seen from Earth

sun dog

sun dog

Mars : Venus mean orbits (diagrammatic)

Above: The two circles we most commonly see around the Sun depict the mean orbits of the only two planets between us and the Sun, a multiple coincidence which only works for observers on Earth. Adapted engraving from Flammarion, 1885.

THE STARRY SIGNATURE
circumstantial evidence for life on earth

***

Despite all the scientific discoveries over recent centuries we are today possibly as far from understanding what we are doing here as the ancients were from being able to build a pocket calculator. The ancients, however, pondered consciousness deeply, and held that the soul was particularly nourished by the applied arts of geometry and music. Through these arts they carefully investigated the relationship between 'the One' and 'the Few', for in music there are only so many notes in tune, and in geometry only so many shapes that fit.

This book has shown simple and beautiful examples of harmony and geometry in the solar system. The Golden Section, long associated with life, and conspicuously absent from modern equations, plays lovingly around Earth. Does this in some way have something to do with 'why we are here', and if so could these techniques be used to locate intelligent life in other solar systems?

I hope you have enjoyed reading *Quadrivium*, and that the cosmos has been beautified or transmuted in some way as a result of what you have learned. If you ever need reminding that there may be a little more magic to the universe than modern cosmology can yet offer, then just remember the kiss of Venus and the words of John Donne:

> *Man hath weav'd out a net, and this net throwne*
> *upon the Heavens, and now they are his owne.*
> *Loth to goe up the Hill, or labour thus*
> *to goe to Heaven, we make Heaven come to us.*

APPENDICES
& INDEX

EARLY NUMBER SYSTEMS

The ancient systems below all use small sets of characters to represent a limited range of numbers. Those from the ancient Mediterranean repeat marks like a tally stick to make some numbers while the Chinese system combines the characters for one to nine with characters meaning 10, 100, 1,000 and 10,000. In all these systems a number such as 57 would be written as the character (or characters) for 50 followed by the character for 7 with no place value.

	Egyptian Heiroglyphs	Egyptian Cursive	Cretan 'Linear B'	Greek Attic (Athens)	Sheba (South Arabia)	Early Roman	Medieval Roman	Archaic Chinese	Chinese Seal Script	Classical Chinese
1	𐤉	J	I	I	I	I	I	—		一
2	𐤉𐤉	U	II	II	II	II	II	=		二
3	𐤉𐤉𐤉	Ш	III	III	III	III	III	☰		三
4	𐤉𐤉𐤉𐤉	ШU	IIII	IIII.	IIII	IIII	IV	☰		四
5	𐤉𐤉	٦	IIIII	Γ	Ψ	V	V	Ⴟ		五
6	𐤉𐤉𐤉	٤	IIIIII	ΓI	ΨI	VI	VI	∧		六
7	𐤉𐤉𐤉𐤉	⌐ᴣ	IIIIIII	ΓII	ΨII	VII	VII	+		七
8	𐤉𐤉𐤉𐤉	≈	IIIIIIII	ΓIII	ΨIII	VIII	VIII	)(		八
9	𐤉𐤉𐤉𐤉	٦	IIIIIIIII	ΓIIII	ΨIIII	VIIII	IX	ᛤ		九
10	∩	ʌ	—	Δ	ο	X	X	—|		一十
20	∩∩	ʌ	=	ΔΔ	οο	XX	XX	=|		二十
30	∩∩∩	X	≡	ΔΔΔ	οοο	XXX	XXX	☰|		三十
40	∩∩∩∩	↵	≣	ΔΔΔΔ	οοοο	XXXX	XL	☰|		四十
50	∩∩∩	٦	≣	Γ⁵⁰	Ρ	Ψ	L	Ⴟ|		五十
60	∩∩∩	ш	==≡	Γ⁵⁰Δ	Ρο	ΨX	LX	∧|		六十
70	∩∩∩∩	ๆ	===≡	Γ⁵⁰ΔΔ	Ροο	ΨXX	LXX	+|		七十
80	∩∩∩∩	Ш	===≡	Γ⁵⁰ΔΔΔ	Ροοο	ΨXXX	LXXX	)(|		八十
90	∩∩∩	๛	==≡	Γ⁵⁰ΔΔΔΔ	Ροοοο	ΨXXXX	XC	ᛤ|		九十
100	𓎆	⌐	Ο	H	Ɓ	✳	C	—◊		一百
500	𓎆𓎆	⌐	Ο⁵⁰⁰	Ⱶ	ƁƁƁƁƁ	◿	D	Ⴟ◊		五百
1,000	𓆼	Ь	◇	X	𓏏	⊠	M	—ᘔ		一千
5,000	𓆼𓆼𓆼 𓆼𓆼	⊔ ⊔	◇◇◇ ◇◇	Γ¹⁰⁰⁰	𓏏𓏏𓏏𓏏𓏏			Ⴟᘔ		五千
10,000	𓂭		⊖	M						一萬

PLACE VALUE NUMBER SYSTEMS

Systems of numerals that use position or 'place value' to signify the magnitude of a given digit are few and far between. The earliest such system is Sumerian cuneiform. Stylus impressions in clay are repeated to make glyphs for 1 to 59, with place value denoting larger numbers. Later the Babylonians introduced an 'empty place' marker effectively making the first zero.

| 1 | 2 | 3 | 4 | 5 | 6 | 7 | 8 | 9 | 10 | 20 | 30 | 40 | 50 | .... | 54 | .... | 59 |

| 2 | 7 | 39 | $2 \times 3,600 + 7 \times 60 + 39 = 7,659$ | | 9 | 0 | 7 | $9 \times 3,600 + 0 \times 60 + 7 = 32,407$ |

The Maya independently discovered place value and the use of zero. Their base 20 system is usually written vertically. Digits in the 3rd place are not 20 but 18 times those in the second, probably because of the calendrical use of 360.

| 1 | 2 | 3 | 4 | 5 | 6 | 7 | 8 | 9 | 10 |

| 11 | 12 | 13 | 14 | 15 | 16 | 17 | 18 | 19 | 0 |

$$\begin{array}{l} 12 \times 360 \\ + \\ 3 \times 20 \\ + \\ 19 \\ \hline 4,399 \end{array}$$

$$\begin{array}{l} 4 \times 7,200 \\ + \\ 17 \times 360 \\ + \\ 6 \times 20 \\ + \\ 0 \\ \hline 35,040 \end{array}$$

Far Eastern rod numerals alternate two versions of nine glyphs. The small Indian zero was adopted in the 18th century.

| 1 | 2 | 3 | 4 | 5 | 6 | 7 | 8 | 9 |

2 1 6 0   2 0 7 3 6

Our own number system originates in Indian Brahmi numerals. From the 6th century onwards variations of the first 9 Brahmi digits were used with a small circular zero in a place value system. This system was passed to Europe by the Arabs.

	1st Century Numerals (from Brahmi)
	8th Century Nagari (Central India)
	10th Century Eastern Arabic (Hindi)
	11th Century Europe (from Ghubar)
	Contemporary Nagari
	Contemporary Arabic (Hindi)
1 2 3 4 5 6 7 8 9 0	Contemporary European/International

357

Pythagorean Numbers

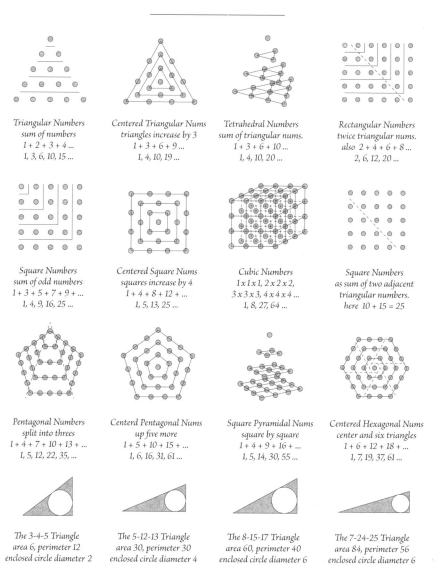

Triangular Numbers
sum of numbers
1 + 2 + 3 + 4 ...
1, 3, 6, 10, 15 ...

Centered Triangular Nums
triangles increase by 3
1 + 3 + 6 + 9 ...
1, 4, 10, 19 ...

Tetrahedral Numbers
sum of triangular nums.
1 + 3 + 6 + 10 ...
1, 4, 10, 20 ...

Rectangular Numbers
twice triangular nums.
also 2 + 4 + 6 + 8 ...
2, 6, 12, 20 ...

Square Numbers
sum of odd numbers
1 + 3 + 5 + 7 + 9 + ...
1, 4, 9, 16, 25 ...

Centered Square Nums
squares increase by 4
1 + 4 + 8 + 12 + ...
1, 5, 13, 25 ...

Cubic Numbers
1 x 1 x 1, 2 x 2 x 2,
3 x 3 x 3, 4 x 4 x 4 ...
1, 8, 27, 64 ...

Square Numbers
as sum of two adjacent
triangular numbers.
here 10 + 15 = 25

Pentagonal Numbers
split into threes
1 + 4 + 7 + 10 + 13 + ...
1, 5, 12, 22, 35, ...

Centerd Pentagonal Nums
up five more
1 + 5 + 10 + 15 + ...
1, 6, 16, 31, 61 ...

Square Pyramidal Nums
square by square
1 + 4 + 9 + 16 + ...
1, 5, 14, 30, 55 ...

Centered Hexagonal Nums
center and six triangles
1 + 6 + 12 + 18 + ...
1, 7, 19, 37, 61 ...

The 3-4-5 Triangle
area 6, perimeter 12
enclosed circle diameter 2

The 5-12-13 Triangle
area 30, perimeter 30
enclosed circle diameter 4

The 8-15-17 Triangle
area 60, perimeter 40
enclosed circle diameter 6

The 7-24-25 Triangle
area 84, perimeter 56
enclosed circle diameter 6

EXAMPLES OF GEMATRIA

Ancient Greek and Christian gematria;

ΙΗΣΟΥΣ + ΧΡΙΣΤΟΣ = 2368
(Jesus) 888 (Christ) 1480

888 : 1480 : 2368 = 3 : 5 : 8

ΚΑΙ Ο ΑΡΙΘΜΟΣ ΑΥΤΟΥ ΧΞΣ = 2368
(And his number is 666)

ΤΟ ΑΓΙΟΝ ΠΝΕΥΜΑ + ΠΑΡΑ ΘΕΟΥ = 1746
(The Holy Spirit) 1080 (from God) 666

Η ΔΟΞΑ ΤΟΥ ΘΕΟΥ ΙΣΡΑΗΛ = 1746
(Glory of The God of Israel)

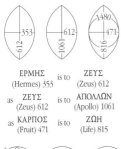

ΕΡΜΗΣ is to ΖΕΥΣ
(Hermes) 353 (Zeus) 612

as ΖΕΥΣ is to ΑΠΟΛΛΩΝ
(Zeus) 612 (Apollo) 1061

as ΚΑΡΠΟΣ is to ΖΩΗ
(Fruit) 471 (Life) 815

ΗΛΙΟΣ (Sun) = 318   ΒΙΟΣ (Life) = 282

1000 as magnified Unity

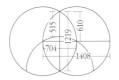

ΠΑΡΘΕΝΟΣ (Virgin) = 515

ΞΥΛΟΝ (Cross) = 610

Ο ΘΕΟΣ ΙΣΡΑΗΛ
(The God of Israel) = 703

ΙΧΘΟΣ (Fish) = 1219

ΣΩΤΗΡ (Saviour) = 1408

The Divine Name YHWH as Tetractys;

```
     י               = 10
    ה  י            = 10 + 5 = 15
   ו  ה  י         = 10 + 5 + 6 = 21
  ה  ו  ה  י      = 10 + 5 + 6 + 5 = 26
```

היה HaYaH (He Was) = 25

הוה HoWeH (He is) = 16

יהיה YiHYeH (He shall be) = 30

Some Hebrew correspondences;

אחד = 13 = אהבה
EKHAD (One)   AHAVAH (Love)

and their sum = 26 = YHWH

אדם חוה = 26 = יהוה
ADAM – KHAWAH   YHWH

יין = 70 = סוד
YAYIN (wine)   SOD (secret)

or *in vino veritas!*

Hebrew letter names and their totals;

אלף	111	ALEPH	למד	74 LAMED
בית	412	BET	מים	90 MEM
גמל	73	GIMMEL	נון	110 NUN
דלת	434	DALET	סמך	120 SAMEKH
הא	6	HE	עין	130 AYIN
וו	12	VOV	פה	85 PE
זין	67	ZAYIN	צדי	104 TSADE
חית	418	HET	קוף	104 QUF
טית	419	TET	רֵישׁ	510 RESH
יוד	20	YOD	שין	360 SHIN
כף	100	KOF	תו	406 TAV

Every number contains the seed of the
next so the rules of gematria allow a
difference of 1 in comparisons.
Fractional parts in geometric measures
and ratios can be rounded either way.

Talisman with magic sum of 66, the abjad total for the Divine Name Allah;

21	26	19
20	22	24
25	18	23

Some names of God in abjad order;

الله	66	ALLAH
باقي	113	BAQI (Everlasting)
جامع	114	JAMI (Gatherer)
ديان	65	DAYAN (Judge)
هادي	20	HADI (Guide)
ولي	46	WALI (Friend)
زكي	37	ZAKI (Purifier)
حق	108	HAQ (Truth)
طاهر	215	TAHIR (Pure)
يسين	130	YASSIN (Chief)
كافي	111	KAFI (Sufficient)
لطيف	129	LATIF (Subtle)
ملك	90	MALIK (King)
نور	256	NUR (Light)
سميع	180	SAMI (All Hearing)
علي	110	'ALI (Most High)
فتاح	489	FATAH (Revealer)
صمد	134	SAMAD (Eternal)
قادر	305	QADIR (Powerful)
رب	202	RAB (Lord)
شفيع	460	SHAFI (Healer)
توب	408	TAWAB (Oft Forgiving)
ثابت	903	THABIT (Stable)
خالق	731	KHALIQ (Creator)
ذاكر	921	DHAKIR (Rememberer)
ضار	1,001	DAR (Chastiser)
ظاهر	1,106	DHAHIR (Apparent)
غفور	1,285	GHAFUR (Forgiving)

FURTHER MAGIC SQUARES

A magic square is *normal* if it uses whole numbers from 1 to the square of its order, and *simple* if its only property is rows, columns and main diagonals adding to the magic sum. The normal magic square of order-3 is unique apart from 8 possible reflections and rotations or *aspects* (*4 below*).

2	7	6
9	5	1
4	3	8

6	7	2
1	5	9
8	3	4

2	9	4
7	5	3
6	1	8

4	9	2
3	5	7
8	1	6

If the numbers in a magic square sum symmetrically about the center, for example 2 + 8, 7 + 3… the square is *associated* (not simple), the number pairs are *complementary*.

There are 880 order-4 normal magic squares. To count magic squares mathematicians rotate/reflect them so the top-left cell is as small as possible with the cell to its right less than the cell below. Complementary numbers in normal order-4 squares form 12 *Dudeney patterns* (*4 shown below*).

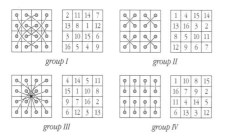

2	11	14	7
13	8	1	12
3	10	15	6
16	5	4	9

group I

1	4	15	14
13	16	3	2
8	5	10	11
12	9	6	7

group II

4	14	5	11
15	1	10	8
9	7	16	2
6	12	3	13

group III

1	10	8	15
16	7	9	2
11	4	14	5
6	13	3	12

group IV

The 48 Group I squares are *pandiagonal*, the 6 broken diagonals formed by opposite sides wrapping round to meet each other also sum magically (*below left and center*).

Order-4 pandiagonal magic squares are also *most-perfect*,

4	14	11	5	4	14
15	1	8	10	15	1
6	12	13	3	6	12
9	7	2	16	9	7
4	14	11	5	4	14
15	1	8	10	15	1

4	14	11	5	4	14
15	1	8	10	15	1
6	12	13	3	6	12
9	7	2	16	9	7
4	14	11	5	4	14
15	1	8	10	15	1

4	14	11	5	4	14
15	1	8	10	15	1
6	12	13	3	6	12
9	7	2	16	9	7
4	14	11	5	4	14
15	1	8	10	15	1

any 2-by-2 square, including wrap arounds, adds to the magic sum (*above right*). Only normal pandiagonal squares of doubly-even order (4, 8, 12…) can be most-perfect.

1	15	24	8	17
23	7	16	5	14
20	4	13	22	6
12	21	10	19	3
9	18	2	11	25

There are 275,305,224 normal order-5 magic squares. Order-5 is the lowest order of magic squares that can be both pandiagonal and associated (*one shown here*). There are 36 essentially different pandiagonal order-5 magic squares, each produces 99 variations by permuting rows, columns and diagonals for a total of 3,600 pandiagonal order-5 squares. It is not known how many normal order-6 magic squares there are. Order-6 is the first oddly-even order, divisible by 2 but not by 4, the hardest squares to construct. It is impossible for a normal order-6 square to be pandiagonal or associated.

1	2	3	4	5	6	7	8
9	10	11	12	13	14	15	16
17	18	19	20	21	22	23	24
25	26	27	28	29	30	31	32
33	34	35	36	37	38	39	40
41	42	43	44	45	46	47	48
49	50	51	52	53	54	55	56
57	58	59	60	61	62	63	64

64	2	3	61	60	6	7	57
9	55	54	12	13	51	50	16
17	47	46	20	21	43	42	24
40	26	27	37	36	30	31	33
32	34	35	29	28	38	39	25
41	23	22	44	45	19	18	48
49	15	14	52	53	11	10	56
8	58	59	5	4	62	63	1

To construct a magic square of doubly-even order, place the numbers in sequence from top left as below. Using the pattern shown exchange every number on a marked diagonal with its complement and you have a magic square.

To make a magic square of any odd order place 1 in the top middle cell and place numbers in sequence up and to the right by one cell, wrapping top/bottom and right/left as necessary. When a previously filled cell is reached move down one cell instead. The central cell will contain the middle number of the sequence and the diagonals will add to the magic sum (*alternative fill pattern below right*).

30	39	48	1	10	19	28
38	47	7	9	18	27	29
46	6	8	17	26	35	37
5	14	16	25	34	36	45
13	15	24	33	42	44	4
21	23	32	41	43	3	12
22	31	40	49	2	11	20

13	23	40	1	18	35	45
21	31	48	9	26	36	4
22	39	7	17	34	44	12
30	47	8	25	42	3	20
38	6	16	33	43	11	28
46	14	24	41	2	19	29
5	15	32	49	10	27	37

Two magic squares combine to make a *composition magic* square with the original orders multiplied together.

1	14	7	12
15	4	9	6
10	5	16	3
8	11	2	13

2	7	6
9	5	1
4	3	8

16	96	80
128	64	0
48	32	112

Make copies of the first square (*left*) as if each were a cell in the second square (*center*). Subtract 1 from each cell in the second square and multiply by the number of cells in the first square (*right*). Add these to each cell in the large square.

17	30	23	28	97	110	103	108	81	94	87	92
31	20	25	22	111	100	105	102	95	84	89	86
26	21	32	19	106	101	112	99	90	85	96	83
24	27	18	29	104	107	98	109	88	91	82	93
129	142	135	140	65	78	71	76	1	14	7	12
143	132	137	134	79	68	73	70	15	4	9	6
138	133	144	131	74	69	80	67	10	5	16	3
136	139	130	141	72	75	66	77	8	11	2	13
49	62	55	60	33	46	39	44	113	126	119	124
63	52	57	54	47	36	41	38	127	116	121	118
58	53	64	51	42	37	48	35	122	117	128	115
56	59	50	61	40	43	34	45	120	123	114	125

To make a *bordered* magic square add double the order, plus 2 to the cells of a normal magic square and make a border of the highest/lowest numbers in the new sequence.

5	4	24	25	7
3	12	17	10	23
18	11	13	15	8
20	16	9	14	6
19	22	2	1	21

14	10	17	6	18
2	11	25	3	24
19	5	13	21	7
22	23	1	15	4
8	16	9	20	12

2	10	19	14	20
22	3	21	11	8
17	25	13	1	9
18	15	5	23	4
6	12	7	16	24

bordered square   *inlaid square*   *inlaid diamond*

A magic square within another that doesn't follow the highest/lowest number border rule is an *inlaid* magic square. Also possible are *inlaid magic diamonds* and *embedded* magic squares (*orders 3 & 4 in order-7 below*).

9	1	37	48	38	26	16
49	10	23	47	4	18	24
15	22	36	11	29	42	20
7	33	44	25	43	17	6
35	46	14	2	21	27	30
19	32	8	3	28	40	45
41	31	13	39	12	5	34

A *bimagic* square is still magic if all its numbers are squared. This one has a magic sum of 369. Each 3-by-3 section also has this sum. The 'squared' magic sum is 20,049.

1	23	18	33	52	38	62	75	67
48	40	35	77	72	55	25	11	6
65	60	79	13	8	21	45	28	50
43	29	51	66	58	80	14	9	19
63	73	68	2	24	16	31	53	39
26	12	4	46	41	36	78	70	56
76	71	57	27	10	5	47	42	34
15	7	20	44	30	49	64	59	81
32	54	37	61	74	69	3	22	17

In 3 dimensions we find the surprising possibility of magic cubes. There are 4 order-3 normal magic cubes (*2 shown below*) each has 48 aspects. All rows, columns, pillars and the four long diagonals from opposite vertices sum to 42.

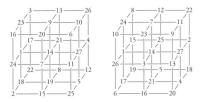

Even more remarkably magic figures in 4-dimensions, once considered impossible, were first discovered by John R. Hendricks who sketched a magic tesseract, or 4-D cube, in 1950. Below is one of 58 normal order-3 magic tesseracts.

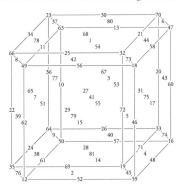

361

Some Numbers of Things

1

One One (*Universal*).

2

Two Forces (*Taoist*): Receptive (*yin*), Active (*yang*).
Two Perspectives (*Universal*): Subject, Object.
Two Polarities (*Geography*): North, South.
Two Polarities (*Physics*): Positive, Negative.
Two Principles (*Metaphysics*): Essence, Substance.
Two Regents (*Alchemy*): Queen, King.
Two Sides Left, Right.
Two Tribal Types (*Anthropology*): Settler, Nomad.
Two Truths (*Logic*): Analytic (*a priori*), Synthetic (*a posteriori*).
Two Ways of Knowing (*Religion*): Esoteric, Exoteric.

3

The Great Triad (*Taoist*): Heaven, Man, Earth.
The Holy Trinity (*Christian*): Father, Son, Holy Ghost.
Three Alchemical Stages (*Alchemy*): Blackening (*nigredo*), Whitening (*albedo*), Reddening (*rubedo*).
Three Aspects of Knowledge (*Greek*): The Knower, Knowing, The Known.
Three Dialectic Phases: Thesis, Antithesis, Synthesis.
Three Dimensions (*Physical*): Medial, Lateral, Vertical.
Three Fates (*Greek*): Spinner (*Clotho*), Measurer (*Lachesis*), Cutter (*Atropos*).
Three Furies (*Greek*): Avenging Murder (*Tisiphone*), Jealousy (*Megaera*), Unceasing Anger (*Alecto*).
Three Generation of Quarks (*Physics*): Up & Down, Charm & strange, Top & Bottom.
Three Graces (*Greek*): Splendour (*Aglaia*), Mirth (*Euphrosyne*), Good Cheer (*Thalia*).
Three Gunas (*Hindu*): Fire (red), Water (white), Earth (black).
Three Kingdoms (*Medieval*): Animal, Vegetable, Mineral.
Three Modes (*Astrology*): Cardinal, Fixed, Mutable.
Three Parts of the Atom (*20th C*): Proton, Neutron, Electron.
Three Parts of a Syllogism (*Aristotle*): Premise, Universal Principle, Conclusion.
Three Primary Colours (*Light*): Red, Green, Blue.
Three Principles (*Alchemy*): Sulphur, Mercury, Salt.
Three Qualities (*Christian*): Faith, Hope, Love.
Three Regular Tilings (*Geometry*): Triangles, Squares, Hexagons.
Three Revolutionary Virtues (*French*): Liberty, Equality, Fraternity.

Three Stages (*Hindu*): Creating (*Brahma*), Sustaining (*Vishnu*), Destroying (*Shiva*).
Three Stages of the Soul (*Jain*): External, Internal, Supreme.

4

Four Beautiful Harmonies (*Music*): Unison, Octave, Fifth, Fourth. All arise from ratios involving the first four numbers.
Four Castes (*Hindu*): Holy (*Brahmin*), Heroic (*Kshatriya*), Business (*Vaisya*), Servant (*Shudra*).
Four Causes (*Aristotle*): Formal, Material, Efficient, Final.
Four Directions (*Common*): North, South, East, West.
Four Elements (*Western*): Fire, Earth, Air, Water.
Four Forces (*Modern*): Electromagnetic, Strong Nuclear, Weak Nuclear, Gravitational.
Four Humours (*Western*): Sanguine, Choleric, Phlegmatic, Melancholic.
Four Levels of Psyche (*Jung*): Ego, Shadow, Anima/Animus, Self.
Four Modes of Pysche (*Jung*): Feeling, Thinking, Sensing, Intuiting.
Four Noble Truths (*Buddhism*): The Truth, Cause, Cessation and Path to cessation of suffering.
Four Seasons (*Western*): Spring, Summer, Autumn, Winter.
Four Types of Literature (*Western*): Romance, Tragedy, Irony, Comedy.

5

Five Animals (*Chinese*): Scaled, Winged, Naked, Furred, Shelled.
Five Directions and Colours (*Chinese*): East (green), South (red), Center (yellow), West (white), North (black).
Five Elements (*Chinese*): Fire, Earth, Metal, Water, Wood.
Five Elements (*Buddhist*): Void, Water, Earth, Fire, Air.
Five Notes (*Chinese*): Black keys on a piano.
Five Orders of Architecture (*Western*): Tuscan, Doric, Ionic, Corinthian, Composite.
Five Parts of the Personality (*Egyptian*): Name, Shade, Life force (*Ka*), Character (*Ba*), Spirit (*Akh = Ka + Ba*).
Five Platonic Solids (*Universal*): Tetrahedron, Octahedron, Cube, Icosahedron, Dodecahedron.
Five Poisons (*Buddhist*): Confusion, Pride, Envy, Hatred, Desire.
Five Precepts (*Buddhist*): Respect for Life, Respect for Property, Chastity, Sobriety, Speaking the Truth.
Five Senses (*Common*): Sight, Hearing, Touch, Smell, Taste.
Five Sounds (*Chinese*): Calling, Laughing, Singing, Lamenting, Moaning.

Five Smells (*Chinese*): Goatish, Burning, Fragrant, Rank, Rotten.

Five Tastes (*Chinese*): Sour, Bitter, Sweet, Spicy, Salty.

Five Virtues (*Buddhist*): Kindness, Goodness, Respect, Economy, Altruism.

Five Virtues (*Chinese*): Benevolence, Propriety, Good Faith, Righteousness, Knowledge.

6

Six Days of Creation (*Abrahamic*): Light, Firmament, Land & Vegetation, Heavenly Bodies, Fish & Birds, Animals & Man.

Six Directions (*Common*): Up, Down, Left, Right, Front, Back.

Six Kingdoms (*Modern*): Archaebacteria & Bacteria (prokaryotes). Protista, Fungi, Plantae & Animalia (eukaryotes).

Six Perfections (*Buddhism*): Giving, Morality, Patience, Energy, Meditation, Wisdom.

Six Reactions (*Chemistry*): Synthesis & Decomposition, Combustion, Single & Double Displacement, Acid-Base.

Six Realms (*Hindu & Buddhist*): Gods, Hells, Human, Hungry Ghosts, Demons, Animals.

Six Regular Polytopes (*4-Dimensional Solids*): Simplex, Tesseract, 16-Cell, 24-Cell, 120 Cell, 600-cell.

7

Seven Border Geometries (*Universal*): There are seven possible kinds of border symmetry.

Seven Chakras (*Hindu*): Root (4 petals), Sacral (6), Solar Plexus (10), Heart (12), Throat (16), Brow (2), Crown (1000).

Seven Deadly Sins and their Seven Contrary Virtues (*Christian*). Humility against pride, Kindness against envy, Abstinence against gluttony, Chastity against lust, Patience against anger, Liberality against envy, Diligence against sloth.

Seven Endocrine Glands (*Medical*): Pineal, Pituitary, Thyroid, Thymus, Adrenal, Pancreas, Gonads.

Seven Heavenly Bodies and their Days (*Ancient*): Moon (Mon), Mercury (Wed), Venus (Fri), Sun (Sun), Mars (Tues), Jupiter (Thurs), Saturn (Sat).

Seven Levels of Self (*Anthroposophy*): Physical, Etheric, Astral, Ego, Manas, Buddhi, Atma.

Seven Liberal Arts (*Western*): Logic, Rhetoric, and Grammar (*Trivium*): Number, Music, Geometry, Cosmology (*Quadrivium*).

Seven Metals (*Ancient*): Silver, Mercury, Copper, Gold, Iron, Tin, Lead.

Seven Modes (*Greek*): Ionian, Dorian, Phrygian, Lydian, Myxolydian, Aeolian, Locrian: using just the white keys on a piano, these refer to the seven-note scale starting with C, D, E, F, G, A and B, respectively.

Seven Stages of the Soul (*Sufi*): Compulsion, Conscience, Inspiration, Tranquility, Submission, Servant, Perfected.

Seven Virtues (*Christian*): Faith, Hope, Charity, Fortitude, Justice, Prudence, Temperance.

8

Eight Semi-Regular Tilings (*Geometry*): In the plane.

Eight Immortals (*Taoist*): Youth, Old Age, Poverty, Wealth, The Populace, Nobility, The Masculine, The Feminine.

Eight Limbs of Yoga (*Vedic*): Morality (*Yama*), Observances (*Niyama*), Postures (*Asanas*), Breathing (*Pranayama*), Concentration (*Dharana*), Devotion (*Dhyana*), Union (*Samadhi*).

Eight Trigrams (*I-Ching*): Chi'en (Heaven, Creative), *Tui* (Attraction, achievement), *Li* (Awareness, beauty), *Chen* (Action, movement), *Sun* (Following, Penetration), *K'an* (Danger, Peril), *Ken* (Stop, Rest), *K'un* (Earth, Receptive).

Eightfold Path (*Buddhist*): Right View, Right Speech, Right Action, Right Livelihood, Right Effort, Right Mindfulness, Right Concentration.

9

Nine Muses (*Greek*): History (*Clio*), Astronomy (*Urania*), Tragedy (*Melpomene*), Comedy (*Thalia*), Dance (*Terpsichore*), Songs to the Gods (*Polyhymnia*), Epic Poetry (*Calliope*), Love Poetry (*Erato*), Lyric Poetry (*Euterpe*).

Nine Orders of Angels (*Western*): Angels, Archangels, Virtues, Powers, Principalities, Dominations, Throne, Cherubim, Seraphim.

Nine Personalities (*Middle-Eastern*): Perfectionist, Giver, Achiever, Tragic Romantic, Observer, Contradictor, Enthusiast, Leader, Mediator.

Nine Regular Polyhedra (*Universal*): The *five Platonic Solids* plus the four stellated polyhedra the great, the stellated, and the great stellated dodecahedra, and the great icosahedron.

Nine Semi-Regular Tilings (*Universal*): Although there are eight standard patterns, one of these has different left- and right-handed versions, making nine in all.

10

Ten Commandments (*Christianity*): Honour Mother, Father & Sabbath. No other gods, Graven images, Blaspheming, Killing, Adultery, Stealing, False witness, Coveting.

Ten Levels (*Buddhism*): Joyous, Stainless, Light-maker, Radiant, Resilient, Turning toward, Far-going, Unshakeable, Good Mind, Cloud of Dharma.

Ten Opposites (*Pythagoras*): Limited:Unlimited, Odd:Even, Singularity:Plurality, Right:Left, Male:Female, Resting:Moving, Straight:Curved, Light:Dark, Good:Bad, Square:Oblong.

Ten Sephiroth (*Kabbalah*): *Kether, Chokmah, Binah, Chesed, Geburah, Tiphareth, Netzach, Hod, Yesod, Malkuth.*

Select Glossary of Number

1   The 1st triangular, square, pentagonal, hexagonal, tetrahedral, octahedral, cubic, Fibonacci, Lucas num.

2   The 1st even (female) num. The planet Mercury's day is exactly two of its years. Uranus orbital radius is two of Saturns. Neptune's period is twice Uranus'.

3   The 1st Greek odd (masculine) num. $1 + 2$. There are three regular tilings of the plane. After three years the Moon closely repeats its phases in the calendar. In engineering triangulation creates stability.

4   The 2nd square num. $2^2 = 2 \times 2 = 2 + 2$. Num of vertices and faces of a tetrahedron. Every integer is the sum of at most 4 squares.

5   Sum of the 1st male and female nums. $1^2 + 2^2$. Five notes in the pentatonic scale. Five Platonic solids. The 5th Fibonacci num and the 2nd pentagonal num.

6   The 3rd triangular num as $6 = 1 + 2 + 3$. The factorial of 3, written $3! = 1 \times 2 \times 3$. Area and semi-perimeter of the 3-4-5 triangle. The first perfect num. (sum of its factors). Edges on a tetrahedron, faces on a cube, vertices of an octahedron. Six regular 4-D polytopes.

7   There are seven frieze symmetries. Seven notes in the traditional scale. Seven endocrine glands in humans. Sum of spots on opposite sides of a dice. Seven tetrominos (Tetris). 4th Lucas num.

8   The 2nd cube, $2^3 = 2 \times 2 \times 2 = 8$. Faces on an octahedron, vertices on a cube. The 6th Fibonacci num. Eight semi-regular tilings. Bits in a Byte.

9   The square of three. $3^2 = 3 \times 3 = 1^3 + 2^3$. There are nine regular polyhedra and nine semi-regular tilings of the plane if you include the chiral pair. In base 10, the digits of all multiples of 9 eventually sum to 9.

10   The 4th triangular and 3rd tetrahedral num. $= 1^2 + 3^2$.

11   11 dimensions unify the four forces of physics. The 5th Lucas num. Sunspot cycle in years.

12   12 notes complete the equal-tempered scale. The 3rd pentagonal num. 12 spheres touch a central one as the cuboctahedron. Num of vertices of an icosahedron, faces of a dodecahedron, edges of both cube and octahedron. Petals of the heart chakra.

13   The 7th Fibonacci num. There are 13 Archimedean polyhedra. Appears as the octave (13th note), and in the 5-12-13 triangle. Locusts swarm every 13 years.

14   The 3rd square pyramidal num $= 1^2 + 2^2 + 3^2$. Num

of lines in a sonnet (octave, quartet, couplet).

15   Triangular num. Sum in lines of a 3×3 magic square. Balls in a snooker triangle.

16   2^4 and 4^2. Perimeter and area of a 4×4 square. Petals of the throat chakra. Also $5^2 - 3^2$ which means 16 coins can be arranged as a square 5 by 5.

17   The number of 2-D symmetry groups. $1^4 + 2^4$. Syllables in Japanese Haiku $(5 + 7 + 5)$. Tones in Arabic tuning.

18   The number of years in a Saros eclipse cycle before you get a eclipse of the same kind near the same place.

19   The number of years in the Metonic cycle. After 19 years full moons recur on the same calendar dates. The game of Go features a 19 by 19 grid.

20   The sum of the first 4 triangular numbers. Num of faces in an icosahedron, vertices in a dodecahedron. Days in a Mayan month. Amino acids in humans.

21   The 6th triangular and 8th Fibonacci num. 3×7. Letters in the Italian alphabet.

22   Max. num. of pieces into which a cake can be cut with 6 slices. Channels in the Kabbalah. Letters in the Hebrew alphabet. Major Arcana in Tarot. Tones in Indian tuning.

23   Chromosome pairs make a human.

24   Spheres can touch one in 4-D. Letters in Greek alphabet. $4! = 1 \times 2 \times 3 \times 4$.

25    $5^2 = 3^2 + 4^2$. 25 raised to any power ends in 25.

26   The only number to sit between a square and a cube. Num of letters in Latin and English alphabets.

27    $3^3 = 3 \times 3 \times 3$. The number of nakshatras into which the ecliptic is divided in Hindu lunar cosmology.

28   The 2nd perfect num, sum of its factors. Triangular. Num of letters in Arabic and Spanish alphabets.

29   A Lucas num, the series goes 1, 3, 4, 7, 11, 18, 29 etc. Letters in Norwegian alphabet.

30   Edges on both dodecahedra and icosahedra. Area and perimeter of a Pythagorean 5-12-13 triangle. The Moon orbits the Earth at a distance of 30 Earth diameters.

31   Planes of existence in Buddhism. A Mersenne prime, of the form $2^n - 1$, where n is prime.

32   2^5. The smallest 5th power besides 1. Num of crystal classes. Num of Earth diameters to reach the Moon.

33    $1! + 2! + 3! + 4!$. Num of vertebrae in the human spinal column, carrying 33 pairs of nerves. 12053

	sunrises in 33 years. Largest num which cannot be represented as sum of distinct triangular nums.
34	The sum in the lines of a 4 × 4 magic square.
35	Sum of Pythagorean harmonic sequence 12:9:8:6. Also the sum of the first five triangular numbers.
36	$1^3 + 2^3 + 3^3$. 8th triangular and 6th square number. First number which is square and triangular.
37	The heart of the $111, 222 \ldots 666, 777, 888$ sequence. 37 moons in 3 years. Stages of the Buddhist bodhisattva.
38	38 can be written as the sum of two odd numbers in 10 different ways. Each pair contains a prime. Largest number with this property.
39	There are 39 hand patterns when a deck of card is divided between four people, as in bridge.
40	The number of fingers and toes of a man and woman together. 40 Spheres can touch one in 5 dimensions.
41	The expression $x^2 - x + 41$ produces a sequence of 40 consecutive primes, from 41 all the way to 1681.
42	The sum in the lines of a 3-D 3 × 3 × 3 magic square.
45	Triangular, sum of 1 to 9. Sum of lines in Sudoku.
46	Total number of chromosomes in human cell nuclei, 23 from mum, 23 from dad.
50	Num of letters in sanscrit alphabet, petals of all chakras excluding crown.
52	Num of playing cards in a pack. Num of human teeth over lifetime (4 × 5) children's + (4 × 8) adult. The Mayan calendar round was 52 years, at which point the 260 day Tzolkin and the 365 day Haab reset.
55	Highest triangular & Fibonacci num (others 1, 3, 21). Also square pyramidal $1 + 4 + 9 + 16 + 25$.
56	Station stones at Stonehenge. Useful for eclipse prediction. 7 × 8. The product of $1 + 2 + 4$ and $1 × 2 × 4$. Tetrahedral. Minor Arcana in Tarot.
58	There is one stellation of a pentagon or hexagon, two of a heptagon or octagon, three for an enneagon. There are no stellations of a tetrahedron or cube, one of an octahedron, three of a dodecahedron ... but there are 58 stellations of an icosahedron.
59	There are two full moons every 59 days. Prime.
60	3 × 4 × 5. Basis of Sumerian and Babylonian counting. Smallest number divisible by 1 through 6.
61	Codons specify amino acids in human mRNA.
64	Eight squared, four cubed and 2^6. Num of hexagrams in the I-Ching, squares on a chess-board. 64 codons specify amino acids in human DNA.
65	The sum in the lines of a 5 × 5 magic square. The first

	number that is the sum of two squares in two ways as $65 = 1^2 + 8^2 = 4^2 + 7^2$.
71	The Hindu Indra lives for 71 eons.
72	Spheres can touch one in 6 dimensions. 360/5. 72 names of God in Kabbalah. 1 lifetime = 1 precessional 'Great Day' or 360th of a Great Year = 72 years. The Rule of 72: How long will it take for my money to double? If interest rate is 6%, then it will take 72/3 = 12 years. You can also use 71 and 70.
73	73 Tzolkin = 52 Haab in the Mayan calendar. 73 is 365/5 and appears in ancient year-clocks.
76	Years between sightings of Halley's comet.
78	Complete Tarot, 22 major and 56 minor Arcana. Triangular, sum of 1 to 12. Number of presents in *12 Days of Christmas*.
81	The square of nine. 3^4. There are 81 stable elements.
89	Fibonacci number common in sunflowers.
91	A quarter of a year, 7 × 13. Square pyrmadial, sum of first 6 squares.
92	Elements can occur in nature; all others appear fleetingly under laboratory conditions.
97	There are 97 leap years every 400 years in the Gregorian calendar Number of cards in the Minchiate Tarot. 78 cards (*see 78*) plus four virtues, four elements and 12 signs of zodiac.
99	Names of Allah. 99 full moons occur in 8 years.
100	10 × 10 in any base.
108	$1^1 × 2^2 × 3^3$. The Sun diameter is 109 times Earth's, and its distance from Earth is 107 Sun diams. Num of Hindu and Buddhist prayer beads.
111	The sum in the lines of a 6 × 6 magic square. Num of Moon diameters between Moon and Earth.
120	1 × 2 × 3 × 4 × 5. Triangular and tetrahedral.
121	The square of eleven.
125	The cube of five.
128	2^7. The largest num not the sum of distinct squares.
144	The square of 12. Only square Fibonacci num.
153	The number of fishes in the net in St. John's Gospel, XX1.11. $= 1^3 + 3^3 + 5^3 = 1! + 2! + 3! + 4! + 5! =$ the square of the number of full moons in a year. Archimedes' approximation for √3 was 265/153
169	The square of thirteen.
175	The sum in the lines of a 7 × 7 magic square.
206	Bones in an adult human body.
216	Plato's nuptial number. The smallest cube that is the sum of three cubes, $6^3 = 3^3 + 4^3 + 5^3$. Twice 108.

219	There are 219 3-D symmetry groups.
220	Member of the smallest amicable pair with 284, the factors of each summing to the other.
235	The number of full moons in a 19-year Metonic cycle.
243	3^5. Appears in the leimma, the Pythagorean halftone 256:243 between the third and fourth notes.
256	2^8. In computers, the maximum value of a byte.
260	The Mayan Tzolkin, $20 \times 13 = 260$ days. Magic sum of 8×8 magic square.
284	Amicable with 220, summing with it to 504.
300	Babies are born with 300 bones.
343	The cube of 7.
354	Days in 12 full moons. Lunar or Islamic year.
360	$3 \times 4 \times 5 \times 6$. Degrees in a circle. Days in a Mayan Tun.
361	The square of 19. A Chinese Go board is 19 by 19.
364	The number of pips on a pack of playing cards, counting J=11, Q=12, K=13. Also = $4 \times 7 \times 13$.
365	The Mayan Haab, consisted of 18 months of 20 days each, plus five days added on (Wayeb) to make 365.
369	Magic sum of 9×9 magic square.
384	Root number for Pythagorean musical scale.
400	The Sun is 400 times larger than the Moon, and 400 times further away.
432	72×6. 108×4. Second note in Pythagorean scale, 9/8 up from 384.
486	Pythagorean major third, two tones up from 384.
496	The third perfect num, sum of its factors.
504	$7 \times 8 \times 9$.
512	2^9. Fourth, 4:3 (or 9/8 × 9/8 × 256/243) up from 384. The cube of 8.
540	There are 540 double doors to Valhalla. Half 1080.
576	Perfect fifth, 3:2 up from 384. 24^2.
584	Venus' synodic period in days. = 8×73.
648	Pythagorean sixth, 3:2 up from the second (432).
666	Sum of numbers 1 to 36. Yang principle in gematria. The sum of the first six Roman numerals (I V X L C D).
720	$6! = 1 \times 2 \times 3 \times 4 \times 5 \times 6 = 8 \times 9 \times 10$. 2×360.
729	Pythagorean seventh, 3:2 up from the third (486). The cube of 9. 3^6 or 27^2. Appears in Plato's *Republic*.
780	Mars' synodic period, in days. = 13×60.
873	$1! + 2! + 3! + 4! + 5! + 6!$.
880	Num of substantially different 4×4 magic squares.
1,000	The cube of 10 in any base.
1,080	$2^3 \times 3^3 \times 5$. Canonical. Yin principle in gematria. Radius of Moon in miles.
1,225	The second triangular and square num. 35^2.
1,331	The cube of 11.
1,461	There are 1461 days in 4 years.
1,540	One of only five triangular AND tetrahedral nums.
1,728	The cube of 12. Cubic inches in a cubic foot.
1,746	Canonical. The sum of 666 and 1080.
2,160	720×3. Canonical. Diameter of Moon in miles. Years in a precessional 'great month' or astrological age.
2,187	3^7.
2,392	= $8 \times 13 \times 23$. The Maya discovered that $3^4 = 81$ full moons occur every 2392 days to astonishing accuracy.
2,920	= $584 \times 5 = 365 \times 8$. The number of days it takes Venus to draw its pentagonal pattern around Earth.
3,168	$2^5 \times 3^2 \times 11$. Canonical. Factors add to 6660.
3,600	The square of 60. Seconds in an hour or degree.
5,040	Radius of Earth in miles. $7! = 1 \times 2 \times 3 \times 4 \times 5 \times 6 \times 7 = 7 \times 8 \times 9 \times 10$. Combined radii of Earth and Moon.
5,913	$1! + 2! + 3! + 4! + 5! + 6! + 7!$
7,140	Largest triangular and tetrahedral num.
7,200	Mayan Katun, or 20 Tuns of 360 days.
7,920	Diameter of Earth in miles. = 720×11.
8,128	The fourth perfect number, sum of its factors.
10,000	A myriad.
20,736	$12 \times 12 \times 12 \times 12$.
25,770	Current value for precession (seems to be slowing, suggesting the Sun forms a binary system with Sirius).
25,920	12×2160. Years in the ancient western count for the precessional cycle of astrological ages.
26,000	Mayan precessional figure.
31,680	Perimeter of a square drawn around earth, in miles.
40,320	$8! = 1 \times 2 \times 3 \times 4 \times 5 \times 6 \times 7 \times 8$.
45,045	The first triangular, pentagonal and hexagonal num.
86,400	Number of seconds in a day.
108,000	One season of a Kali Yuga.
142,857	The repeating part of all divisions by seven.
144,000	Days in a Mayan Baktun of 20 Katuns.
248,832	12^5.
362,880	$9!$, also = $2! \times 3! \times 3! \times 7!$
365,242	Days in 1000 years. Feet in one equatorial degree.
432,000	The Hindu final and corrupt Kali Yuga period, in years.
864,000	The Hindu third phase of creation, the semi-corrupt dwapara yuga, in years.
1,296,000	The Hindu secondary Treta Yuga period, in years. = 3×432000.
1,728,000	The Hindu initiatory and highly spiritual Satya Yuga period, in years. = 4×432000.
1,872,000	Years in the Mayan long count (ends Dec 2012).
3,628,800	$10!$, also $6! \times 7!$, or $3! \times 5! \times 7!$
4,320,000	The Hindu Mahayuga, a complete cycle of Yugas, a Buddhist Kalpa.
39,916,800	$11!$, also 5040×7920.

Further Numbers

	Triangular	Square	Pentagonal	Centered Triangular	Centered Square	Centered Pentagonal	Rectangular	Tetrahedral	Octahedral	Cubic	Centered Cube	Square Pyramidal	Fibonacci	Lucas
1	1	1	1	1	1	1	1	1	1	1	1	1	1	1
2	3	4	5	4	5	5	2	4	6	8	9	5	1	3
3	6	9	12	10	13	13	6	10	19	27	35	14	2	4
4	10	16	22	19	25	25	12	20	44	64	91	30	3	7
5	15	25	35	31	41	41	20	35	85	125	189	55	5	11
6	21	36	51	46	61	61	30	56	146	216	341	91	8	18
7	28	49	70	64	85	85	42	84	231	343	559	140	13	29
8	36	64	92	85	113	113	56	120	344	512	855	204	21	47
9	45	81	117	109	145	145	72	165	489	729	1241	285	34	76
10	55	100	145	136	181	181	90	220	670	1000	1729	385	55	123
11	66	121	176	166	221	221	110	286	891	1331	2331	506	89	199
12	78	144	210	199	265	265	132	364	1156	1728	3059	650	144	322
13	91	169	247	235	313	313	156	455	1469	2197	3925	819	233	521
14	105	196	287	274	365	365	182	560	1834	2744	4941	1015	377	843
15	120	225	330	316	421	421	210	680	2255	3375	6119	1240	610	1364
16	136	256	376	361	481	481	240	816	2736	4096	7471	1496	987	2207
17	153	289	425	409	545	545	272	969	3281	4913	9009	1785	1597	3571
18	171	324	477	460	613	613	306	1140	3894	5832	10745	2109	2584	5778
19	190	361	532	514	685	685	342	1330	4579	6859	12691	2470	4181	9349
20	210	400	590	571	761	761	380	1540	5340	8000	14859	2870	6765	15127
21	231	441	651	631	841	841	420	1771	6181	9261	17261	3311	10946	24476
22	253	484	715	694	925	925	462	2024	7106	10648	19909	3795	17711	39603
23	276	529	782	760	1013	1013	506	2300	8119	12167	22815	4324	28657	64079
24	300	576	852	829	1105	1105	552	2600	9224	13824	25991	4900	46368	103682
25	325	625	925	901	1201	1201	600	2925	10425	15625	29449	5525	75025	167761
26	351	676	1001	976	1301	1301	650	3276	11726	17576	33201	6201	121393	271443
27	378	729	1080	1054	1405	1405	702	3654	13131	19683	37259	6930	196418	439204
28	406	784	1162	1135	1513	1513	756	4060	14644	21952	41635	7714	317811	710647
29	435	841	1247	1219	1625	1625	812	4495	16269	24389	46341	8555	514229	1149851
30	465	900	1335	1306	1741	1741	870	4960	18010	27000	51389	9455	832040	1860498
21	496	961	1426	1396	1861	1861	930	5456	19871	29791	56791	10416	1346269	3010349
32	528	1024	1520	1489	1985	1985	992	5984	21856	32768	62559	11440	2178309	4870847
33	561	1089	1617	1585	2113	2113	1056	6545	23969	35937	68705	12529	3524578	7881196
34	595	1156	1717	1684	2245	2245	1122	7140	26214	39304	75241	13685	5702887	12752043
35	630	1225	1820	1786	2381	2381	1190	7770	28595	42875	82179	14910	9227465	20633239
36	666	1296	1926	1891	2521	2521	1260	8436	31116	46656	89531	16206	14930352	33385282

Primes

2 3 5 7 11 13 17 19 23 29 31 37 43 47 53 59 61 67 71 73 79 83 89 97 101 103 107 109 113 127 131 137 139 149 151 157 163 167 173 179 181 191 193 197 199 211 223 227 229 233 239 241 251 257 263 269 271 277 281 283 293 307 311 313 317 331 337 347 349 353 359 367 373 379 383 389 397 401 409 419 431 433 439 443 449 457 461 463 467 479 487 491 499 503 509 521 523 541 547 557 563 559 571 577 587 593 599 601 607 613 617 619 631 641 643 647 653 659 661 673 677 683 691 701 709 719 727 733 739 743 751 757 761 769 773 787 797 809 811 821 823 827 829 839 853 857 859 863 877 881 883 887 907 911 919 929 937 941 947 953 967 971 977 983 991 997 1009

RULER & COMPASS CONSTRUCTIONS

The small selection of constructions shown here, taken from *Ruler & Compass* by Andrew Sutton, are given to assist the keen student of Sacred Geometry. They use a simple code. *Line AB* means *draw the straight line that passes through A and B*. *Segment* is used in place of *line* for the section of a straight line defined by two endpoints. *Circle O-A* means *draw a circle centered at O and passing through A*.

Circle radius AB center O means *draw a circle of compass opening length AB centered at O*. *Arc* is used in place of *circle* for drawing only part of the circle. Sometimes, extra points are given to help improve accuracy when drawing, for example, *line ACB*, or *circle O-AB*. Newly found points are noted in brackets. Occasionally a line made possible by new points is assumed drawn, and merely noted.

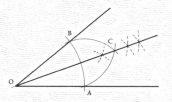

Bisecting an angle:
1. Arc center O (A, B); 2. Arcs A-B, B-A
(C), alternatives shown dashed; 3. Line OC;
∠AOC = ∠BOC = ¹/₂ ∠AOB

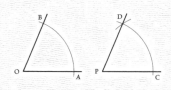

Copying an angle on a given line:
1. Arc center O (A, B); 2. Arc radius OA
center P (C); 3. Arc radius AB center C (D);
4. Line PD; ∠CPD = ∠AOB

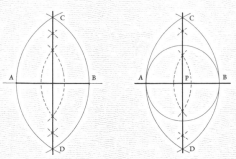

Perpendicular bisector
on a given segment AB:
1. Arcs of equal radius
centers A, B (C, D);
2. Line CD

Perpendicular through a
point P on a line:
1. Circle center P (A, B);
2. Arcs of equal radius centers
A, B (C, D); 3. Line CPD

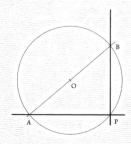

Perpendicular through
a point P on a line:
1. For any point O not on the line,
circle O-P (A); 2. Line AO (B);
Line PB is perpendicular to line AP

368

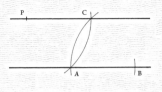

Parallel through a given point P:
1. Arc any suitable radius center P (A);
2. Arc same radius center A (B);
3. Arc same radius center B (C);
Line PC is parallel to line AB

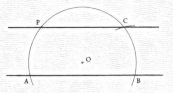

Parallel line through a given point P:
1. Arc O-P any suitable center O (A, B);
2. Arc radius AP center B (C);
Line PC is parallel to line AB

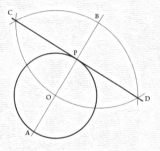

Tangent to a circle:
1. Line OP (A); 2. Arc radius PA
center O (B); 3. Arc B-O (line CPD);
Line CPD is tangent to circle at P

Parallel line at a given distance:
1. Arcs radius equal to given distance,
centers any two points A, B on the
line; 2. Line touching arcs as
shown is parallel to AB

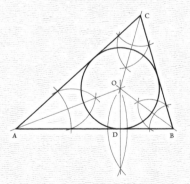

Incenter & incircle of a triangle:
1. Bisect ∠CAB, ∠ABC, ∠BCA (O);
2. Perp. to BC through O (D); 3. Circle O-D

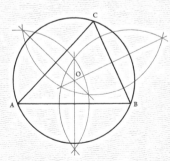

Circumcenter & circumcircle of a triangle:
1. Perp. bisectors on AB, BC, CA (O);
2. Circle O-ABC

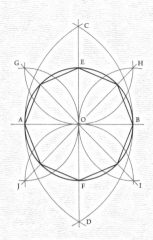

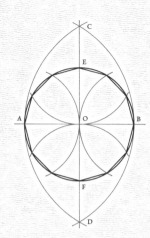

Regular octagon in a circle:
1. Line through center O (A, B); 2. Arcs A-B,
B-A (line CEFD); 3. Arcs A-O, E-O, B-O,
F-O (G, H, I, J); 4. Lines GI, HJ & complete

Regular dodecagon in a circle:
1. Line through center O (A, B);
2. Arcs A-B, B-A (line CEFD);
3. Arcs A-O, E-O, B-O, F-O & complete

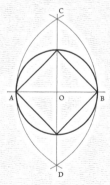

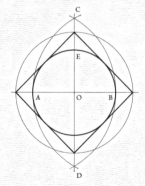

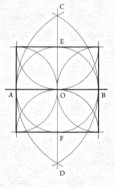

Square in a circle:
1. Line through center O
(A, B); 2. Arcs A-B, B-A;
3. Line CD & complete

Square around a circle:
1. Line through center O (A, B);
2. Arcs A-B, B-A (line CED);
3. Circle radius AE center
O & complete

Square set orthogonally
on a given line:
1. Circle center O on the line
(A, B); 2. Arcs A-B, B-A
(line CEFD); 3. Arcs A-O,
B-O, E-O, F-O & complete

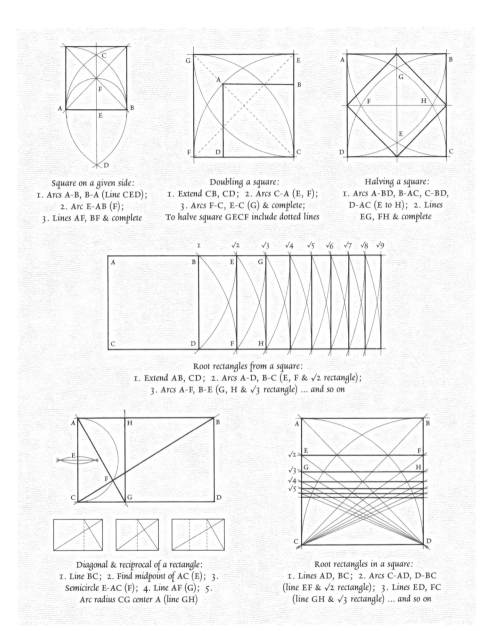

Square on a given side:
1. Arcs A-B, B-A (Line CED);
2. Arc E-AB (F);
3. Lines AF, BF & complete

Doubling a square:
1. Extend CB, CD; 2. Arcs C-A (E, F);
3. Arcs F-C, E-C (G) & complete;
To halve square GECF include dotted lines

Halving a square:
1. Arcs A-BD, B-AC, C-BD,
D-AC (E to H); 2. Lines
EG, FH & complete

Root rectangles from a square:
1. Extend AB, CD; 2. Arcs A-D, B-C (E, F & √2 rectangle);
3. Arcs A-F, B-E (G, H & √3 rectangle) ... and so on

Diagonal & reciprocal of a rectangle:
1. Line BC; 2. Find midpoint of AC (E); 3.
Semicircle E-AC (F); 4. Line AF (G); 5.
Arc radius CG center A (line GH)

Root rectangles in a square:
1. Lines AD, BC; 2. Arcs C-AD, D-BC
(line EF & √2 rectangle); 3. Lines ED, FC
(line GH & √3 rectangle) ... and so on

FLAT-PACKED POLYHEDRA

If a polyhedron is 'undone' along some of its edges and folded flat, the result is known as its net. The earliest known examples of polyhedra presented this way are found in Albrecht Dürer's *Painter's Manual*, from 1525. The nets below are scaled such that if refolded the resulting polyhedra would all have equal circumspheres.

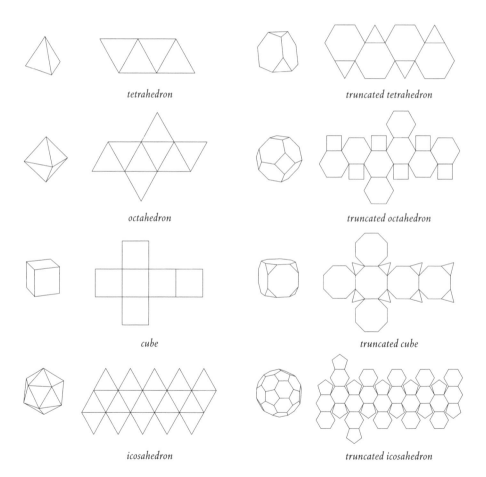

tetrahedron

truncated tetrahedron

octahedron

truncated octahedron

cube

truncated cube

icosahedron

truncated icosahedron

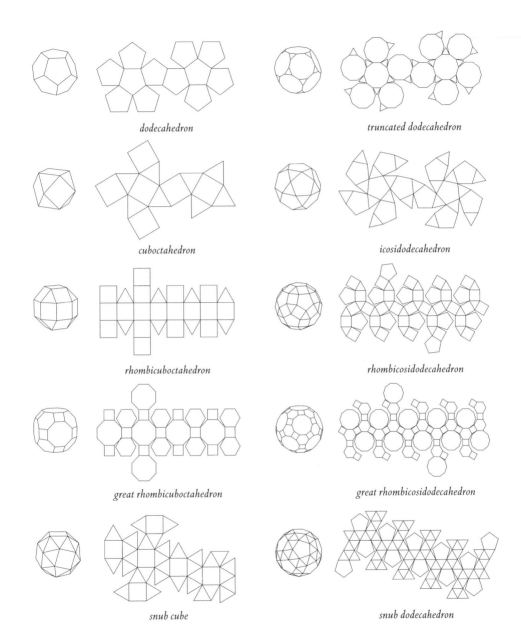

dodecahedron

truncated dodecahedron

cuboctahedron

icosidodecahedron

rhombicuboctahedron

rhombicosidodecahedron

great rhombicuboctahedron

great rhombicosidodecahedron

snub cube

snub dodecahedron

ARCHIMEDEAN SYMMETRIES

The diagrams below show the rotation symmetries of the Archimedean Solids and the two Rhombic Archimedean Duals.

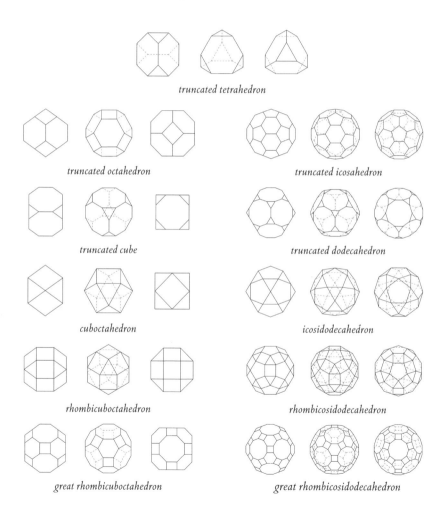

truncated tetrahedron

truncated octahedron

truncated icosahedron

truncated cube

truncated dodecahedron

cuboctahedron

icosidodecahedron

rhombicuboctahedron

rhombicosidodecahedron

great rhombicuboctahedron

great rhombicosidodecahedron

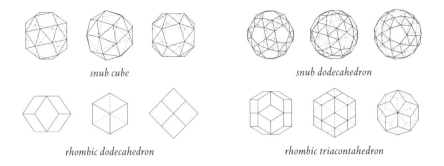

snub cube   *snub dodecahedron*

rhombic dodecahedron   *rhombic triacontahedron*

THREE-DIMENSIONAL TESSELATIONS

Of the Platonic Solids only the cube can fill space with copies of itself and leave no gaps. The only other purely 'Platonic' space filling combines tetrahedra and octahedra. One Archimedean Solid, the truncated octahedron, and one Archimedean Dual, the rhombic dodecahedron, are also space filling polyhedra.

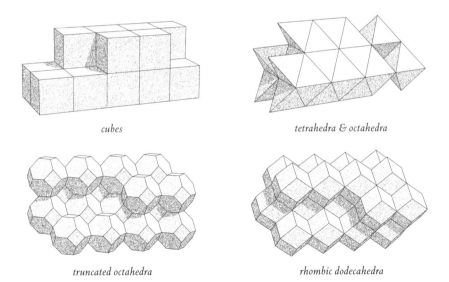

cubes   *tetrahedra & octahedra*

truncated octahedra   *rhombic dodecahedra*

375

Each Embracing Every Other

PLATONIC SOLIDS FORMULÆ

A recurring theme in the metric properties of the Platonic Solids is the occurrence of the irrational numbers Φ (the Golden Section), and the square roots $\sqrt{2}$, $\sqrt{3}$, and $\sqrt{5}$. They are surprisingly elegant when expressed as continued fractions;

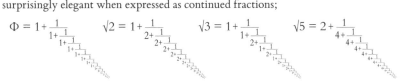

$$\Phi = 1+\cfrac{1}{1+\cfrac{1}{1+\cfrac{1}{1+\cfrac{1}{1+\cdots}}}} \qquad \sqrt{2} = 1+\cfrac{1}{2+\cfrac{1}{2+\cfrac{1}{2+\cfrac{1}{2+\cdots}}}} \qquad \sqrt{3} = 1+\cfrac{1}{1+\cfrac{1}{2+\cfrac{1}{1+\cfrac{1}{2+\cdots}}}} \qquad \sqrt{5} = 2+\cfrac{1}{4+\cfrac{1}{4+\cfrac{1}{4+\cfrac{1}{4+\cdots}}}}$$

Their decimal expansions to twelve places, together with that of π are;

$$\Phi = 1.618033988750 \quad \sqrt{2} = 1.414213562373 \quad \sqrt{3} = 1.732050807569$$
$$\sqrt{5} = 2.236067977500 \quad \pi = 3.141592653590$$

The table below gives volumes and surface areas for a sphere radius r, and Platonic Solids edge length s. Also included are the proportional pathways joining each vertex to every other in the Platonic Solids.

	Volume	Surface Area	Number of Pathways, Length
Sphere	$\frac{4}{3}\pi r^3$	$4\pi r^2$	n/a
Tetrahedron	$\frac{\sqrt{2}}{12}s^3$	$\sqrt{3}\,s^2$	6 edges, s
Octahedron	$\frac{\sqrt{2}}{3}s^3$	$2\sqrt{3}\,s^2$	12 edges, s 3 axial diagonals, $\sqrt{2}\,s$
Cube	s^3	$6s^2$	12 edges, s 12 face diagonals (inscribed tetrahedra), $\sqrt{2}\,s$ 4 axial diagonals, $\sqrt{3}\,s$
Icosahedron	$\frac{5}{6}\Phi^2 s^3$	$5\sqrt{3}\,s^2$	30 edges, s 30 face diagonals, Φs 6 axial diagonals, $\sqrt{(\Phi^2+1)}s$
Dodecahedron	$\frac{\sqrt{5}}{2}\Phi^4 s^3$	$3\sqrt{(25+10\sqrt{5})}s^2$	30 edges, s 60 face diagonals (inscribed cubes), Φs 60 interior diagonals (inscr. tetrahedra), $\sqrt{2}\,\Phi s$ 30 interior diagonals, $\Phi^2 s$ 10 axial diagonals, $\sqrt{3}\,\Phi s$

Polyhedra Data Table

	Symmetry *	Vertices	Edges	Faces (total)	Faces (types)
Tetrahedron	Tetr.	4	6	4	4 triangles
Cube	Oct.	8	12	6	6 squares
Octahedron	Oct.	6	12	8	8 triangles
Dodecahedron	Icos.	20	30	12	12 pentagons
Icosahedron	Icos.	12	30	20	20 triangles
Stellated Dodecahedron	Icos.	12	30	12	12 pentagrams
Great Dodecahedron	Icos.	12	30	12	12 pentagons
Great Stellated Dodecahedron	Icos.	20	30	12	12 pentagrams
Great Icosahedron	Icos.	12	30	20	20 triangles
Cuboctahedron	Oct.	12	24	14	8 triangles 6 squares
Icosidodecahedron	Icos.	30	60	32	20 triangles 12 pentagons
Truncated Tetrahedron	Tetr.	12	18	8	4 triangles 4 hexagons
Truncated Cube	Oct.	24	36	14	8 triangles 6 octagons
Truncated Octahedron	Oct.	24	36	14	6 squares 8 hexagons
Truncated Dodecahedron	Icos.	60	90	32	20 triangles 12 decagons
Truncated Icosahedron	Icos.	60	90	32	12 pentagons 20 hexagons
Rhombicuboctahedron	Oct.	24	48	26	8 triangles 18 squares
Great Rhombicuboctahedron	Oct.	48	72	26	12 squares 8 hexagons 6 octagons
Rhombicosidodecahedron	Icos.	60	120	62	20 triangles 30 squares 12 pentagons
Great Rhombicosidodecahedron	Icos.	120	180	62	30 squares 20 hexagons 12 decagons
Snub Cube	Oct.-**	24	60	38	32 triangles 6 squares
Snub Dodecahedron	Icos.-**	60	150	92	80 triangles 12 pentagons

* Symmetries: Tetrahedral: 4 x 3-fold axes, 3 x 2-fold, 6 mirror planes. Octahedral: 3 x 4-fold axes, 4 x 3-fold, 6 x 2-fold, 9 mirror planes. Icosahedral: 6 x 5-fold axes, 10 x 3-fold, 15 x 2-fold, 15 mirror planes.

** The snub solids have no mirror planes.

Inradius *** Circumradius	Midradius *** Circumradius	Edge Length *** Circumradius	Dihedral Angles ****	Central Angle ****
0.3333333333	0.5773502692	1.6329931619	70°31'44"	109°28'16"
0.5773502692	0.8164965809	1.1547005384	90°00'00"	70°31'44"
0.5773502692	0.7071067812	1.4142135624	109°28'16"	90°00'00"
0.7946544723	0.9341723590	0.7136441795	116°33'54"	41°48'37"
0.7946544723	0.8506508084	1.0514622242	138°11'23"	63°26'06"
0.4472135955	0.5257311121	1.7013016167	116°33'54"	116°33'54"
0.4472135955	0.8506508084	1.0514622242	63°26'06"	63°26'06"
0.1875924741	0.3568220898	1.8683447179	63°26'06"	138°11'23"
0.1875924741	0.5257311121	1.7013016167	41°48'37"	116°33'54"
0.8164965809 0.7071067812	0.8660254038	1.0000000000	125°15'52"	60°00'00"
0.9341723590 0.8506508084	0.9510565163	0.6180339887	142°37'21"	36°00'00"
0.8703882798 0.5222329679	0.9045340337	0.8528028654	70°31'44" 109°28'16"	50°28'44"
0.9458621650 0.6785983445	0.9596829823	0.5621692754	90°00'00" 125°15'52"	32°39'00"
0.8944271910 0.7745966692	0.9486832981	0.6324555320	109°28'16" 125°15'52"	36°52'12"
0.9809163757 0.8385051474	0.9857219193	0.3367628118	116°33'54" 142°37'21"	19°23'15"
0.9392336205 0.9149583817	0.9794320855	0.4035482123	138°11'23" 142°37'21"	23°16'53"
0.9108680249 0.8628562095	0.9339488311	0.7148134887	135°00'00" 144°44'08"	41°52'55"
0.9523198087 0.9021230715 0.8259425910	0.9764509762	0.4314788105	125°15'52" 135°00'00" 144°44'08"	24°55'04"
0.9659953695 0.9485360199 0.9245941063	0.9746077624	0.4478379596	148°16'57" 159°05'41"	25°52'43"
0.9825566436 0.9647979663 0.9049441875	0.9913166895	0.2629921751	142°37'21" 148°16'57" 159°05'41"	15°06'44"
0.9029870683 0.8503402074	0.9281913780	0.7442063312	142°59'00" 153°14'05"	43°41'27"
0.9634723304 0.9188614921	0.9727328506	0.4638568806	152°55'48" 164°10'31"	26°49'17"

*** From the polyhedron's center, the inradius is measured to the various face-centers, the midradius to the edge midpoints, & the circumradius to vertices.

**** In Archimedean Solids the larger dihedral angles are found between smaller pairs of faces.

***** The central angle is the angle formed at the center of a polyhedron by joining the ends of an edge to that center.

Some Featured Tunings

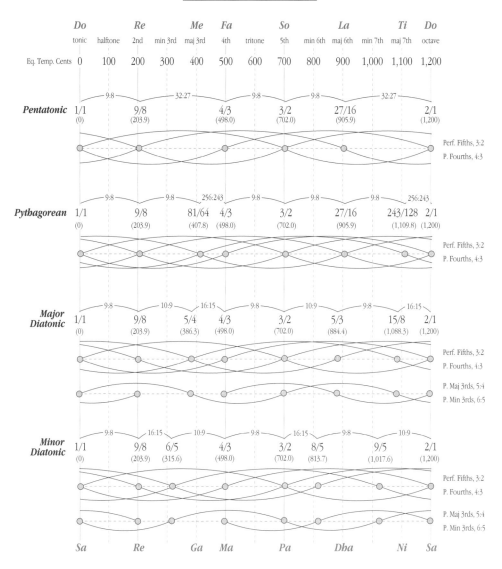

SELECTED MUSICAL INTERVALS

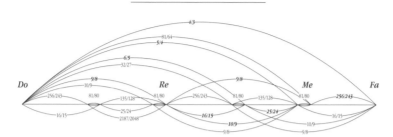

Interval	Cents	Name
1:1	0	Unison
32805:32768	2.0	Schisma
2048:2025	19.6	Diaschisma
81:80	21.5	Syntonic Comma
531441:524288	23.5	Pythagorean Comma
128:125	41.1	Diesis
25:24	70.7	Minor Diatonic Halftone
256:243	90.2	Leimma, Pythag. Halftone
135:128	92.2	Major Chroma
16:15	111.7	Major Diatonic Halftone
2187:2048	113.7	Apotome
27:25	133.2	Large Leimma
10:9	182.4	Minor Tone
9:8	203.9	Major Tone
8:7	231.2	Septimal Tone
7:6	266.9	Septimal Minor Third
32:27	294.1	Pythag. Minor Third
6:5	315.6	Perfect Minor Third
5:4	386.3	Perfect Major Third
81:64	407.8	Pythag. Major Third
4:3	498.0	Perfect Fourth
7:5	582.5	Septimal Tritone
45:32	590.2	Diatonic Tritone
729:512	611.7	Pythag. Tritone
3:2	702.0	Perfect Fifth
128:81	792.2	Pythag. Minor Sixth
8:5	813.7	Diatonic Minor Sixth
5:3	884.4	Perfect Major Sixth
27:16	905.9	Pythag. Major Sixth
7:4	968.8	Harmonic Seventh
16:9	996.1	Pythag. Minor Seventh
9:5	1,017.6	Diatonic Minor Seventh
15:8	1,088.3	Diatonic Major Seventh
243:128	1,109.8	Pythag. Major Seventh
2:1	1,200	Octave

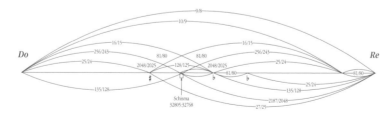

Like the off-center division of the octave into fifths and fourths, sharps are not flats, giving rise to five more notes, making seventeen in all (found in middle-eastern tunings). More completely, we may think of the seven notes of the scale as moving across twelve 'regions' of the octave, falling into the twenty-two positions of Indian tuning.

THE GREEK MODES

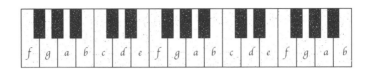

Modern Names	The Seven Modes of Antiquity								Ancient Greek Names
Ionian *Major*	do re mi fa so la ti do	c	d	e f	g	a	b c	1 2 3 4 5 6 7 8	Lydian
Dorian	re mi fa so la ti do re	d	e f	g	a	b c	d	1 2 3♭ 4 5 6 7♭ 8	Phrygian
Phrygian	mi fa so la ti do re mi	e f	g	a	b c	d	e	1 2♭ 3♭ 4 5 6♭ 7♭ 8	Dorian
Lydian	fa so la ti do re mi fa	f	g	a	b c	d	e f	1 2 3 4♯ 5 6 7 8	Syntolydian
Myxolydian	so la ti do re mi fa so	g	a	b c	d	e f	g	1 2 3 4 5 6 7♭ 8	Ionian
Aeolian *Natural Minor*	la ti do re mi fa so la	a	b c	d	e f	g	a	1 2 3♭ 4 5 6♭ 7♭ 8	Aeolian
Locrian	ti do re mi fa so la ti	b c	d	e f	g	a	b	1 2♭ 3♭ 4 5♭ 6♭ 7♭ 2	Myxolydian

The white notes on a piano give the seven notes of the seven modes of ancient Greece. Medieval transcription errors have left us with modern names which don't fit the ancient ones. Each mode, or scale, has its own pattern of tones and halftones, only two surviving as our major and natural minor scales.

Other scales include modal pentatonics which forbid halftones, the harmonic minor with its minor 3rd and 6th, 1 2 3♭ 4 5 6♭ 7 8, and many others.

HARMONIC CONSTANTS AND EQUATIONS

The ratios and intervals in this book concern frequencies, normally expressed as cycles per second, or Hertz. Classical tuning sets C at 256 Hz. Modern tuning is higher, fixing A at 440 Hz. The period T of a wave is the reciprocal of its frequency f: $T = 1/f$.

The speed of sound in dry air is roughly $331.4 + 0.6T_c$ m/s, where T_c is the temperature in degrees celsius. Its value at room temperature, $20°c$, is 343.4 m/s.

Gravitational acceleration on earth, g, is 9.807 m/s².

Frequency of a Pendulum.	$\dfrac{1}{2\pi} \sqrt{\dfrac{gravitational\ acceleration}{pendulum\ length}}$
Fundamental frequency of a tensioned string.	$\dfrac{1}{2 \times string\ length} \sqrt{\dfrac{string\ tension}{string\ mass \div string\ length}}$
Resonant frequency of a cavity with an opening.	$\dfrac{speed\ of\ sound}{2\pi} \sqrt{\dfrac{area\ of\ opening}{volume\ of\ cavity \times length\ of\ opening}}$
Fundamental frequency of an open pipe or cylinder.	$\dfrac{speed\ of\ sound}{2 \times length\ of\ cylinder}$

The beat frequency between f_1 and f_2 is the difference between them, $f_b = f_2 - f_1$.

The ratio $a:b$ converts to cents (where $a > b$): $(\log(a) - \log(b)) \times (1200 \div \log2)$. To convert cents into degrees multiply by 0.3.

Clapping in front of a rise of steps produces a series of echoes with a perceived frequency equal to $v/2d$, where v is the speed of sound, and d is the depth of each step. Clapping in a small corridor width w produces a frequency v/w.

The *arithmetic* and *harmonic means* are central to Pythagorean number theory. The arithmetic mean of two frequencies separated by an octave produces the fifth between them (3:2), the harmonic mean producing the fourth (4:3).

$$6 \quad : \quad 8 \quad :: \quad 9 \quad : \quad 12$$

$$A \quad : \quad \frac{2AB}{A+B} \quad :: \quad \frac{A+B}{2} \quad : \quad B$$

$$A \quad : \quad \underset{Mean}{Harmonic} \quad :: \quad \underset{Mean}{Arithmetic} \quad : \quad B$$

TABLES OF HARMONOGRAPH PATTERNS

Overtone and simple ratio harmonics are shown below and opposite, arranged in order of increasing dissonance down the page. Open phase drawings display their ratio as the number of loops counted across and down. To find the ratio of a rotary drawing, draw both forms, concurrent (both circles in the same direction) and contrary (in opposite directions). Count the loops in each, add the two numbers together and divide the total by two. This gives the larger ratio number. Subtract this from the contrary total to give the lower ratio number. Rotary figures for the ratio $a:b$ will have $b-a$ loops when both circles are concurrent, and $a+b$ loops when they are contrary. The designs shown here were all made with equal amplitudes.

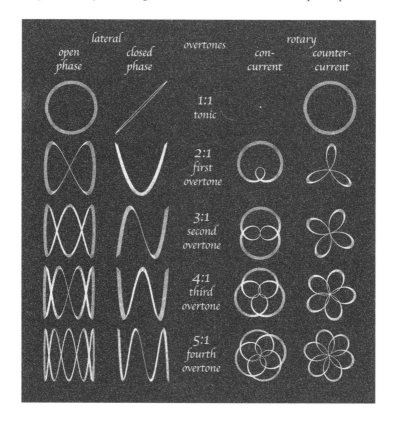

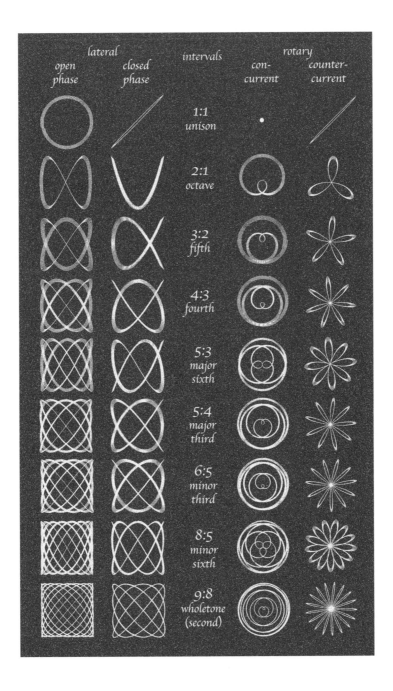

lateral		intervals	rotary	
open phase	closed phase		con-current	counter-current
		1:1 unison		
		2:1 octave		
		3:2 fifth		
		4:3 fourth		
		5:3 major sixth		
		5:4 major third		
		6:5 minor third		
		8:5 minor sixth		
		9:8 wholetone (second)		

BUILDING A HARMONOGRAPH

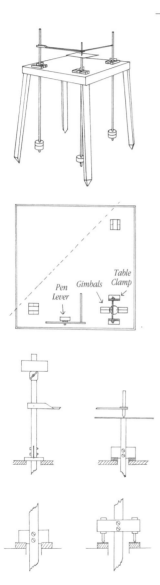

Pen Lever · Gimbals · Table Clamp

ANYONE seriously interested in making a harmonograph should consider going straight for the three-pendulum model.

The table must be highly rigid and firm on the floor, otherwise the movements of the weights will be distorted. I suggest it should be about 90 cm above the floor, 60 x 30 cm for two pendulums, 60 x 60 cm for three, and some 2 cm thick with an apron all round, about 8 cm deep.

Legs should be about 6 cm square, splayed outwards and pointed at the bottom. One way of achieving the splay is to fix wood or metal brackets in the corners under the table each side of the diagonals and bolt the legs between them. After adjusting the legs to give the correct splay they can then be fixed in position with screws through the apron.

To save space, slice off the table as along the dotted line. Three legs are not quite so stable, but work fairly well.

The platform carrying the paper should be light and rigid, fixed to the pendulum with a countersunk screw. A size of 22 x 16 cm will conveniently take half an A4 sheet secured by a rubber band or small clip.

All sizes suggested are maxima, but a scaled down version will still work if it is carefully made.

If you are tempted to make a harmonograph, start with the weights, for the instrument will only be satisfactory if these are really heavy and yet easy to adjust. It is a good idea to make about ten of, say, two kilos each, so the loadings can be varied. They should be about 8cm in diameter, with a central hole, or better with a slot for easier handling. Either cast them yourself from lead or ready-mixed cement or have them made by a metal smallware dealer or friendly plumber.

The shafts should be made from wood dowel, about 1.5 cm in diameter (metal rods are liable to bend, distorting the drawings), marked off in cm.

Clamps can be obtained from suppliers of laboratory equipment. For some of the drawings top weights are needed, held in place by clamps. Clamps can also be added to pendulum tops for 'fine tuning', with one or more metal washers added.

The simpler kind of bearing consists of brass strips bolted into a slot in the pendulum and filed to sharp edges to rest in grooves each side.

In a bearing involving less friction the pendulum is encased at the fulcrum in a horizontal block of hardwood with vertical bolts each side filed to sharp points and resting in dints in metal plates. If drilling the large hole in the block is too difficult, it can be made in two halves, each hollowed out to take the shaft and bolted together.

Rotary motion needs gimbals. Here the grooves for the pendulum are filed in the upper side of a ring (e.g. a key-ring) while the under side

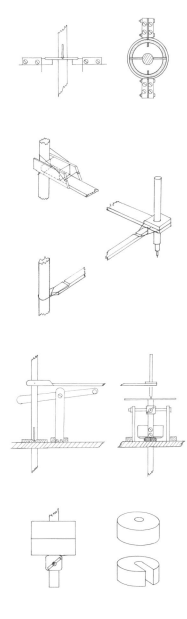

has grooves at rightangles to the upper ones. The lower grooves fit on two projecting sharp edges (brass strips), each enclosed between two pieces of wood fixed to the table. With the alternative bearing a large flat washer should be used with depressions to take the sharp points.

Pen arms should be as light as possible to minimize 'top-hamper'. They are easily made from balsa-wood strips sold by model-making shops, using balsa cement and scotch tape. For two pendulums the arm can be fastened to the shaft with pinched-off needles, and the pen jammed into a hole at the other end. For three pendulums the side pieces on the arm should enclose its shaft firmly but not too tightly and be held gently with a thin rubber band. One of the arms holds the pen, while the other is held by protruding needles pushed in backwards and secured (gently) at both ends by the rubber bands.

There may well be better, more sophisticated ways of doing all this. All suggestions welcomed.

An additional fitting is needed to lock a rotary pendulum so that the instrument can be used with just the two single-axis pendulums. This can be done by mounting two brackets on the table near the rotary pendulum with holes to take a long horizontal bolt (slightly to one side) to which the shaft can be clamped.

Pens should be fine, light and free-flowing. Most stationers and shops selling draughtsmen's and artists' materials offer a variety (avoid ball or thick fibre pens). For best results use shiny 'art' or 'imitation art' paper, ordinary copy-paper for preliminary experiments.

If the pen is left on the paper to the end there is usually an unsightly blob. To avoid this, mount a short pillar on the table with an adjustable lever carrying a piece of thin dowel placed under the pen arm. By raising the dowel gently the pen is lifted off the paper without jogging it. This device should also be used before the pen is lowered to the paper. By watching the pen you can see what pattern is being made, and nudge it one way or the other by pressure on the pendulums.

For ratios outside the octave, such as 4:1, you may need to try another harmonograph such as Goold's twin-elliptic pendulum (right).

387

GLOSSARY OF MUSICAL TERMS

Accidental - Any of the five symbols (bb, b, nat, #, x) that lower or raise a pitch by one or two halftones, usually used to alter or restore a key.

Add - An intermediary or non-chord tone added to a chord for flavouring or colour, usually a 2nd, 4th, 6th, 9th, 11th, or 13th.

Anacrusis - A pickup or upbeat, preceding a metrically strong downbeat.

Appoggiatura - A dissonant tone that occurs on a strong beat, and then resolves to a consonance or chord tone, it 'leans' against the consonance and then relaxes into it.

Augmented - Used to describe an interval or chord. With intervals, it indicates a perfect or major interval raised one halftone, with chords, it indicates a major triad with a raised fifth.

Binary Form - A basic A-B structure in musical form, often each contrasting section repeats.

Cadence - A gesture or assembly of notes and rhythms that suggest a sense of closure, pause, or finality to a musical phrase or section.

Caesura - A pause or rest.

Cambiata - A dissonance formula that is a double neighbour group, effectively a tone and its two flanking tones, above and below.

Chord - Three or more tones sounding together as an independent entity. They are often spelled largely in thirds, the primary core being a triad.

Chromatic - colourful, used to indicate music using halftones, accidentals, or the entire 12-tone collection, contrasted with diatonic.

Clef - A symbol placed at the beginning of the staff that indicates where pitches are to be placed on the lines and spaces of the staff, the three types being the G-clef, F-clef, and C-clef.

Consonance - The relative stability of a musical interval, generally not requiring resolution, contrasted with dissonance. Most often octaves, fifths, fourths, thirds, and sixths.

Counterpoint - The simultaneity of independent lines, which are coherent horizontally and vertically, adhering to strict rules about consonance and dissonance, with historical variance.

Crescendo - A gradual increasing in volume and intensity, indicated by an expanding hairpin in musical notation.

Decrescendo - A gradual decreasing in volume and intensity, indicated by a contracting hairpin in musical notation.

Diatonic - A scale comprised of seven different contiguous tones, with a specific relationship of whole and half steps, often used to describe major and minor. Contrasted with chromatic. Also describes music that adheres to this scale.

Diminished - Used to describe a musical interval or chord, either the lowering of a perfect interval by one halftone, or a triad with a minor third and diminished fifth.

Dissonance - One or more musical intervals that suggest instability, most often seconds and sevenths, and tend to require some form of resolution to a consonance.

Dominant - The scale degree a fifth above a root or tonic of a key, also a powerful station of the scale, that often suggests its own resolution, back to the tonic. Often it is present in a cadence.

Dronal - Music that is primarily melodic and rhythmic, lacking harmonic movement, with all scale tones relating to a drone or still point.

Epigram - As used in this book, a musical motive or idea that has just enough particularity and individuality to constitute a recognizable shape and identity, often functioning as a basic building block of an entire composition.

Éschappée - Escape tone, a dissonance on a weak part of the beat that is approached by step, followed by a leap to a consonance in the opposite direction.

Extension - Tones above the octave added to a chord to enrich its overall colour without altering its function, always a 9th, 11th, or 13th. Also known as a 'tension.'

Flat - A symbol used to indicate a lowering of a natural by one halftone. Also used to describe a tone that is tuned slightly under pitch.

Forte - A symbol in dynamics used to indicate music to be played 'strongly' or loudly, contrasted with piano.

Fricative - A friction-based consonant in spoken language, such as F, V, H, and TH.

Guide Tone - Usually the 3rd and 7th of a chord, acting as leading tones that maximize the forward motion of a harmonic progression.

Half Step - See Halftone.

Harmonic Minor - One of the three types of minor scales, in which the 6th scale degree is lowered and the 7th scale degree is raised, to allow for the dominant chord to be major. Thus it has the interval of an augmented second between the 6th and 7th.

Harmonics - Each of the component tones of the overtones series, which imbue the lowest tone, the fundamental, with timbre or colour, each being mathematically related to it in whole-number ratios, 1x, 2x, 3x, gradually increasing in frequency. Many string and wind instruments can be played in such a way to reveal these softer upper tones.

Harmony - The relationship of a vertical arrangement of tones when sounding together, also a chord, and the way in which chords relate and are organized through time.

Ictus - A metric accent or strong beat, often a downbeat.

Interval - The distance between two pitches.

Inversion - With musical intervals, it is the operation of taking the bottom note and placing it at the top, so a 2nd yields a 7th, 3rd a 6th, 4th a 5th, and so on. In harmony, it indicates that a note other than the root of the chord is in the bass.

Key - A collection of pitches that reinforce one note as a tonic. There are 12 major and 12 minor keys in Western music, each built upon one of the 12 notes, with sharps and flats added accordingly to preserve whole steps and half steps.

Key Signature - The global instruction indicating which notes are to be raised or lowered in a composition to preserve and express a particular scale, placed at the beginning of a staff before the meter and after the clef.

Leading Tone - The seventh scale degree of any scale, most often a half step below the tonic.

Legato - A musical instruction to play the notes in a connected and smooth fashion, with no breaks in between.

Major - In describing musical intervals, this indicates the 2nd, 3rd, 6th, and 7th as they occur naturally in the major scale. In describing chords, a triad that is made up of a perfect fifth, and in between a major third, placed above the root. Contrasted with minor. Also describes an overall scale or key flavour, always referring to the third of the scale.

Measure/Bar - A parcel of musical time, segmented by the meter, in which one complete grouping is delimited. In notation, a measure is separated by a line on either side.

Mediant - The third scale degree of any scale.

Melismatic - Vocal music that has two or more pitches assigned to one syllable.

Melodic Minor - One of the three minors, in this scale the 6th and 7th scale degrees are altered to appear as they do

in the major scale to heighten the upward movement to the tonic. Often a descending version also exists as an unaltered version of the natural minor.

Melody - The succession of tones in time, arranged in a meaningful pattern, which can be of varying lengths.

Meter - A pattern of rhythmic groupings indicated by a fraction, in which the numerator indicates the number of beats per measure, and the denominator indicates the type of subdivisions to receive the beat (quarter, eighth, sixteenth). The two basic forms of meter are duple and triple.

Minor - In describing musical intervals, this indicates a major interval lowered by a halftone or half step. In harmony, a chord that is made up of a perfect fifth, and in between a minor third, placed above the root. Contrasted with major. Also describes an overall scale or key flavour. Western music identifies three types of minors, natural, harmonic, and melodic.

Modal - Music that utilizes scales other than major and minor, such as Phrygian, Dorian, Lydian, etc. (see pp. 8–9). Often this type of music does not modulate.

Modulation - A changing of key or scale in which the tonal center moves, and the accidentals required for one scale are introduced to alter the previous one. This is most easily conveyed by a I-IV-V-I harmonic formula.

Natural - An accidental that cancels a sharp, flat, double sharp, or double flat, corresponding to the white notes on the piano.

Natural Minor - The Aeolian mode, this is the minor scale without any alterations to its 6th or 7th scale degrees.

Neighbour Tone - A non-chord tone that exists above or below a chord tone as a temporary dissonance and decoration, usually on a weak beat.

Octatonic - An eight-note scale, most often referring to diminished scales.

Overtone Series - A natural acoustic phenomenon occurring wherever a vibration of a string or air through a pipe occurs. The length of the vibration increasingly subdivides, yielding vibrations or frequencies higher in sound than the largest vibration. These arrange together to form timbre, and communicate the identity of the sound. Also responsible for the vowels in spoken language.

Passing Tone - An intermediary tone between two chord tones, usually a dissonance, occurring on a weak beat.

Pentatonic - A five-note scale, most often referring to the first five fifths when arranged together, 1-2-3-5-6. Also the black notes on the piano.

Perfect - A musical interval of an octave, 5th, 4th, (and unison), which most closely resemble the first acoustic intervals of the overtone series.

Piano - A musical instruction indicating music to be played 'softly' or quietly. Contrasted with forte.

Plosives - An explosive sounding consonant in spoken language, often a B, P, D, G, or Q.

Polyrhythms - The simultaneous use of two different rhythmic patterns that do not directly relate to one-another, also called 'cross rhythms.'

Register - A specific region of the entire pitch range of an instrument, voice, or piece of music.

Rhythm - The temporal arrangement of movement, quite often possessing a pulse.

Root - The bottommost pitch of a triad or chord, which conveys its function and identity in the context of a harmonic progression. This is always discoverable by arranging the notes into closely voiced thirds.

Rounded Binary - A musical structure in the form ABA, where the last A is a shorter, truncated version of the first.

Halftone - A half step, the smallest interval in Western traditional music. Two adjacent notes on the piano, whether white or black.

Sharp - A symbol used to indicate a raising of a natural by one halftone. Also used to describe a tone that is tuned slightly above pitch.

Sibilants - A type of fricative consonant in spoken language, at higher frequencies, often resembling a hissing sound, such as S, and Z.

Sonata - As used in this book, specifically a formal procedure which presents two contrasting themes, the second being in a key other than the tonic (often the dominant), followed by a modulatory and free development section, and then a recapitulation in which the two themes are presented again, and the second theme is restored to the tonic key.

Staccato - A musical instruction indicating notes to be played in a detached fashion, usually marked by a small dot above or below the note.

Subdominant - The fourth scale degree, a fifth below the tonic, and a whole step below the dominant. It often functions as a departure from the tonic and a preparation for the dominant.

Submediant - The sixth scale degree of any scale, a third below the tonic.

Subtonic - The seventh scale degree of any scale, most often a whole tone below the tonic (see Leading tone).

Supertonic - The second scale degree of any scale, a step above the tonic.

Suspension - A chord in which the third is held or suspended from resolving as the rest of the tones resolve, prepared in the previous chord. In modern music, the suspension needs no preparation, nor resolution.

Syllabic - Vocal music that has only one pitch assigned to a syllable.

Syncopation - A rhythmic procedure by which strong beats are shifted to weak beats, temporarily obscuring the sense of pulse or meter. The use of syncopation is one of the strongest indicators of musical styles around the world.

Tempo - The rate or speed of a musical performance through time.

Tenuto - A musical instruction indicating notes that are to be played in a sustained or extended fashion, usually marked by a line or dash above or below the note.

Ternary - A musical structure in the form ABA, where the last A is generally a complete repetition of the first.

Tonal - Music using the principles of tonic-dominant relationships, predominantly the major and minor systems prevalent in Western music.

Tonic - The home or resting point of a scale, the strongest point of gravity to which all other tones in the scale relate, the first scale degree.

Tonicization - A temporary support of and pointing to a tonal center other than the tonic, but not firmly confirmed, which only occurs in a modulation.

Transposition - A moving of the notes of a composition up or down from one key to another, while keeping the relative intervals intact.

Triad - A three-note structure, arranged in thirds, and resulting in one of the four chord types: major, minor, diminished, or augmented.

Tritone - A diminished 5th, or augmented 4th, made of three whole tones, or six halftones. The largest symmetrical interval in the Western scale, it divides the octave perfectly in half, and is its own inversion. Often utilized in a dominant 7th chord (represented by the fourth and seventh scale degrees of a major scale), it has a powerful forward drive toward resolution by contraction or expansion, and is in fact the liberating force that unlocks the puzzle of tonality by pushing and pulling into and out of one key into another.

Whole Tone - Also a whole step, comprised of two half steps or halftones.

Solfege & Mnemonics

Solfege/Solmization

Ut Re Mi Fa Sol La Si Ut
Gamma-Ut = Gamut

Ut queant laxis, Re-sonare fibris, Mi-ra gestorum
Fa-muli tuorom, Sol-ve polluti, La-bii reatum,
Sa-ncte Iohannes

(8th century plainsong to John the Baptist)

In the West, three systems have evolved and are in use today:

Fixed Do:
Do Re Mi Fa So La Si Do
Advantages: good for those with perfect pitch,
absolute scale names
Disadvantages: any syllable can have up to 5 sounds
(bb, b, nat, #, x)
seven different key syllabifications

Movable Do, chromatic syllables:
Do Di Re Ri Mi Fa Fi So Si La Li Ti Do ascending
Do Ti Te La Le So Se Fa Mi Me Re Ra Do descending
Advantages: Illustrates tonic-dominant Do-So
relationships clearly through major and parallel minor,
one syllabification for all scales
Disadvantages: Obscures relative minor relationship, breaks
down with chromatic passages; i.e. no aug3, no dim4

Movable Do, La minor, chromatic syllables:
Major:
Do Di Re Ri Mi Fa Fi So Si La Li Ti Do ascending
Do Ti Te La Le So Se Fa Mi Me Re Ra Do descending
Minor:
La Li Ti Do Di Re Ri Mi Fa Fi So Si La ascending
La Le So Se Fa Mi Me Re Ra Do Ti Te La descending
Advantages: Reveals the modal and historical
relativity of major and minor
Disadvantages: Masks tonic-dominant Do-So relationship,
breaks down in minor chromatic passages; i.e. no aug5, dim3,
seven different mode syllabifications

Mnemonics

For Thirds:

Treble Clef, lines:
Every Good Boy Does Fine
Every Good Boy Deserves Fudge
Elephants Got Big Dirty Feet
Empty Garbage Before Dad Flips

Treble Clef, spaces:
F-A-C-E, Fun Always Comes Easy

Bass Clef, lines:
Great Big Dandelions Fly Away
Good Boys Do Fine Always
Good Boys Deserve Fudge Always
Great Big Deer from Alaska
Great Big Dogs from America
Granny's Boots Don't Fit Aunty

Bass Clef, spaces:
All Cars Eat Gas
All Cows Eat Grass
All Children Eat Gum

for Fifths and Fourths:

Give Dorothy An Easter Basket For Christmas
For Christmas Give Dorothy An Easter Basket
Flying Birds Enjoy A Delightful Green Countryside
Birds Enjoy A Delightful Green Countryside Flying

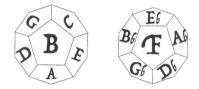

A dodecahedral mapping of the twelve notes of
the scale which preserves the tritone oppositions
shown in the circle of fifths.

Basic Musical Notation

Staff – the lattice or matrix that holds notes and rests in place

Treble Clef – also known as G clef, its inner spiral encircles G

Bass Clef – also known as F clef, its inner spiral encircles F

Alto Clef – also known as C clef, wherever the loops meet is middle C

Tenor Clef – also known as C clef, middle C is a 3rd higher than Alto Clef

Percussion Clef – for rhythmic notation, non-pitched instruments

Ledger Lines – lines drawn above and below the staff to extend the pitch range

Coda – a signpost indicating an ending, placed in the score, and at the end

Segno – a signpost indicating a return, often after the beginning, as a D.S. (Da Segno)

Repeat – like parenthesis, indicating start and end of repeated music

Repeat – repeat previous bar

Fermata – placed over a note or rest, indicating time stopping momentarily

Trill – an ornament, oscillating repeatedly with the note above or below

Trill – an ornament, oscillating once or twice with the note above or below

Up Bow – for strings, an upward bow direction

Down Bow – for strings, a downward bow direction

Martellato – (very marked) a strong accent or punctuation

Staccato – a brief, detached articulation

Harmonic – indicates an overtone to be played in place of the fundamental

Accent/Marcato – a general demarcation, not as strong as martellato

Tenuto – an indication to maximize and connect the duration of the note

8va/8vb – play notes an octave higher or lower than written note

15va/15vb – play notes two octaves higher or lower than written note

stem	**Dotted Note** – adds half the value of the note to the note
	Double Dot – adds half the value of the note to the note, and then half of that
	Arpeggiando – the chord is to be rolled
	Grace Note – smaller, unmeasured decorative note, can come in groups
	Tremolo – indicates a note repeating or vibrating rapidly
beam	**Slur** – legato phrasing, connected notes. For strings: one bow direction. For winds and brass: one breath
notehead	**Tie** – unites the durations of two notes together into one across a beat or barline
1. 2.	**1st and 2nd Endings** – these are form indications that provide for alternate endings for repeated music
/	**Rhythm Slash** – often used in lead sheets, indicates the flow of beats
¢	**Cut Time/Alla Breve** – indicates 2/2

pp	Pianissimo, very soft
p	Piano, soft
mp	Mezzo-piano, medium soft
mf	Mezzo-forte, medium strong
f	Forte, strong, loud
ff	Fortissimo, very strong
sfz	Sforzando – a forceful accent
	Crescendo – increasingly louder
	Decrescendo – increasingly softer
Ped. ✲	**Pedal** – used for piano notation, indicates to depress and release the sustain pedal
	Double Bar (final) – indicates the ending of a score
	Double Bar – indicates the ending of a section
C	**Common Time** – indicates 4/4

sixty-fourth	thirty-second	sixteenth	eighth	quarter	half	whole	US
hemidemisemiquaver	demisemiquaver	semiquaver	quaver	crotchet	minim	semibreve	UK

MUSICAL SCALES

Double Harmonic Minor
1 b2 3 4 5 b6 7 8

Hungarian Major
1 #2 3 #4 5 6 b7 8

Hungarian Minor
1 2 b3 #4 5 b6 7 8

Raga Todi
1 b2 b3 #4 5 b6 7 8

Raga Marva
1 b2 3 #4 5 6 7 8

Blues Scale
1 b3 4 #4 5 b7 8

Persian
1 b2 3 4 b5 b6 7 8

Enigmatic
1 b2 3 #4 #5 #6 7 8

Lydian Minor
1 2 3 #4 5 b6 b7 8

Major Locrian
1 2 3 4 b5 b6 b7 8

Neapolitan Major
1 b2 b3 4 5 6 7 8

Neapolitan Minor
1 b2 b3 4 5 b6 7 8

Mixolydian Augmented
1 2 3 4 #5 6 b7 8

Oriental
1 b2 3 4 b5 6 b7 8

Prometheus
1 2 3 b5 6 b7 8

Chinese
1 3 #4 5 7 8

Leading Whole Tone
1 2 3 #4 #5 #6 7 8

Balinese Pelog
1 b2 b3 4 5 b6 8

Japanese
1 b2 4 5 b6 8

Major Pentatonic
1 2 3 5 6 8

Minor Pentatonic
1 b3 4 5 b7 8

Whole Tone
1 2 3 #4 #5 b7 8

Half-Whole/Diminished/Octatonic
1 b2 b3 ♮3 #4 5 6 b7

Whole-Half/Diminished/Octatonic
1 2 b3 4 b5 b6 ♮6 7

Minor 3rd-half step
1 #2 3 5 b6 8

Half step/minor 3rd
1 b2 3 4 #5 6 8

SELECTED RHYTHMS

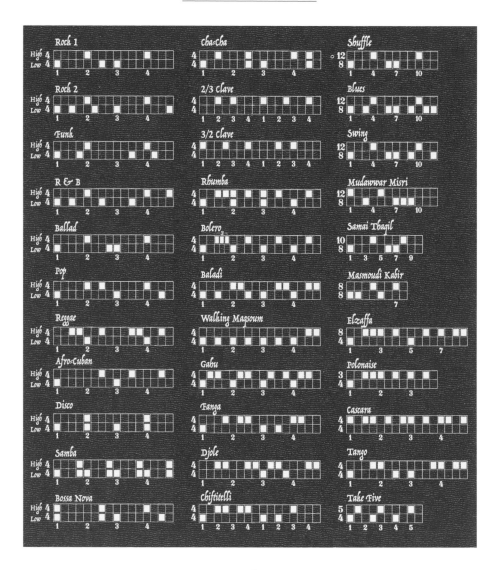

Harmonies

Sun & Planets

		Perihelion $(10^6\ km)$	Mean Orbit $(10^6\ km)$	Aphelion $(10^6\ km)$	Eccentricity	Inclination of Orbit (degrees)	Perihelion Longitude (degrees)	Orbital Period (days)	Tropical Year (days)
The Sun	☉	-	-	-	-	-	-	-	-
Mercury	☿	46.00	57.91	69.82	0.205631	7.0049	77.456	87.969	87.968
Venus	♀	107.48	108.21	108.94	0.006773	3.3947	131.53	224.701	224.695
The Earth	+	147.09	149.60	152.10	0.016710	0	102.95	365.256	365.242
Mars	♂	206.62	227.92	249.23	0.093412	1.8506	336.04	686.980	686.973
Ceres	⚳	446.60	413.94	381.28	0.0789	10.58	???	1680.1	1679.5
Jupiter	♃	740.52	778.57	816.62	0.048393	1.3053	14.753	4,332.6	4,330.6
Saturn	♄	1,352.2	1,433.5	1,514.5	0.054151	2.4845	92.432	10,759.2	10,746.9
Chiron	⚷	1,266.2	2,050.1	2,833.9	0.38316	6.9352	339.58	18,518	18,512
Uranus	♅	2,741.3	2,872.46	3,003.6	0.047168	0.76986	170.96	30,685	30,589
Neptune	♆	4,444.4	4,495.1	4,545.7	0.0085859	1.7692	44.971	60,190	59,800
Pluto	♇	4,435.0	5,869.7	7,304.3	0.24881	17.142	224.07	90,465	90,588

Moons
(a selection)

		Name of Satellite	Mean Orbital Radius $(10^3\ km)$	Orbital Period (days)	Eccentricity of Orbit	Inclination of Orbit (°)	Diameter (mean) (km)	Mass $(10^{18}\ kg)$
Earth's	+	The Moon	384.8	27.3217	0.0549	5.145	3,475	73,490
Mars'	♂	Phobos	9,378	0.31891	0.0151	1.08	22.4	0.0106
		Deimos	23,459	1.26244	0.0005	1.79	12.2	0.0024
Jupiter's	♃	Io	421.6	1.7691	0.004	0.04	3,643	89,330
		Europa	670.9	3.5512	0.009	0.47	3,130	47,970
		Ganymede	1,070	7.1546	0.002	0.21	5,268	148,200
		Callisto	1,883	16.689	0.007	0.51	4,806	107,600
Saturn's	♄	Tethys	294.66	1.8878	<0.001	1.86	1,060	622
		Dione	377.40	2.7369	0.0022	0.02	1,120	1,100
		Rhea	527.04	4.5175	0.0010	0.35	1,528	2,310
		Titan	1,221.8	15.945	0.33	0.33	5,150	134,550
		Iapetus	3,561.3	79.330	0.0283	14.7	1,436	1,590

Rotation Period (hours)	Average Day Length (hours)	Equatorial Diameter (km)	Polar Diameter (km)	Axial Tilt (degrees)	Mass $(10^{24}\ kg)$	Volume $(10^{12}\ km^3)$	Surface Gravity (m/s^2)	Surface Pressure (bars)	Temp. (mean) (^{o}C)
600 - 816	–	1,392,000	1,392,000	7.25	1,989,100	1,412,000	274.0	0.000868	5505
1407.6	4222.6	4,879.4	4,879.4	0.01	0.3302	0.06083	3.70	negl.	167
-5832.5	280.20	12,103.6	12,103.6	177.36	4.8685	0.92843	8.87	92	464
23.934	24.000	12,756.2	12,713.6	23.45	5.9736	1.08321	9.78	1.014	15
24.623	24.660	6794	6750	25.19	0.64185	0.16318	3.69	0.007	-65
9.0744	9.0864	960	932	var.	0.00087	0.000443	negl.	negl.	-90
9.9250	9.9259	142,984	133,708	3.13	1,898.6	1,431.28	23.12	100+	-110
10.656	10.656	120,536	108,728	26.73	568.46	827.13	8.96	100+	-140
5.8992	5.8992	208	148	???	0.000006	0.000024	negl.	negl.	???
-17.239	17.239	51,118	49,946	97.77	86.832	68.33	8.69	100+	-195
16.11	16.11	49,528	48,682	28.32	102.43	62.54	11.00	100+	-215
-153.29	153.28	2390	2390	122.53	0.0125	0.00715	0.58	negl.	-223

MOONS
(continued)

		Name of Satellite	Mean Radius Orbit $(10^3\ km)$	Orbital Period (days)	Eccentricity of Orbit	Inclination of Orbit $(^{o})$	Diameter (mean) (km)	Mass $(10^{18}\ kg)$
Uranus'	♅	Miranda	129.39	1.4135	0.0027	4.22	235.7	66
		Ariel	191.02	2.5204	0.0034	0.31	578.9	1,340
		Umbriel	266.30	4.1442	0.0050	0.36	584.7	1,170
		Titania	435.91	8.7059	0.0022	0.14	788.9	3,520
		Oberon	583.52	13.463	0.0008	0.10	761.4	3,010
Neptune's	♆	Proteus	117.65	1.1223	0.0004	0.55	193	3
		Triton	354.76	-5.8769	0.000016	157.35	2,705	21,470
		Nereid	5,5413	360.14	0.7512	7.23	340	20
Pluto's	♇	Charon	19.6	6.3873	<0.001	<0.01	1,186	1,900

Only the major moons of the gas giants are given. In 2001 there were 28 known moons around Jupiter, 30 around Saturn, 21 around Uranus and 8 around Neptune. There are probably many more. There are 29.5306 days between full moons on Earth. Cosmology can seriously improve your health.

Planetary Tunings

	Mercury	Venus	Earth	Mars	Jupiter	Saturn	Uranus	Neptune	Pluto	Moon	Moon Sid	Day
Period (Years)	0.24085	0.6152	1	1.8809	11.862	29.458	84.013	164.79	247.69	0.08085	0.07480	0.00274
Audible Freq. Hz	141.27	221.23	136.10	144.72	183.58	147.85	207.36	211.44	140.25	210.42	227.43	194.18
Octave No.	30	32	32	33	36	37	39	40	40	29	29	24
Audible Tone	D	A	C♯	D	F♯	D	G♯	A	C♯	G♯	A♯	G
Tuning Pitch Hz	423.34	442.46	432.10	433.67	436.62	443.04	439.37	422.87	445.26	445.86	429.33	435.92
Visible Freq. 10^{14} Hz	6.213	4.865	5.986	6.365	4.037	6.502	4.559	4.650	6.168	4.627	5.001	4.270
Light Oct. No.	72	73	74	75	77	79	80	81	82	70	70	65
Wavelength	0.483	0.616	0.501	0.471	0.743	0.461	0.685	0.645	0.486	0.648	0.599	0.702
Color	blue	orange	blue-green	blue	red	blue	orange-red	orange-red	blue	orange-red	yell-orange	orange-red

adapted from The Cosmic Octave by Hans Cousto

Table of Ancient Measures

	English feet	Assyrian	Iberian	Roman	Common Egyptian	English	Greek	Persian	Belgic	Sumerian	English Archaic	Royal Egyptian	Russian
Assyrian	0.9	1	63/64	15/16		9/10	7/8	6/7					
Iberian	0.91429	64/63	1	20/21	14/15		8/9			5/6		4/5	
Roman	0.96	16/15	21/20	1	49/50	24/25	14/15			7/8			
Common Egyptian	0.97959		15/14	50/49	1	48/49	20/21					6/7	
English	1	10/9		25/24	49/48	1	35/36	20/21	14/15		9/10	7/8	6/7
Greek	1.02857	8/7	9/8	15/14	21/20	36/35	1	48/49	24/25	15/16		9/10	
Persian	1.05	7/6				21/20	49/48	1	49/50				9/10
Belgic	1.07143					15/14	25/24	50/49	1		27/28	15/16	
Sumerian	1.09714		6/5	8/7			16/15			1		24/25	
English Archaic	1.11111					10/9			28/27		1	35/36	20/21
Royal Egyptian	1.14286		5/4		7/6	8/7	10/9		16/15	25/24	36/35	1	48/49
Russian	1.16667					7/6		10/9			21/20	49/48	1

Ancient Metrology

Geometry means 'Earth measure' and the measures of antiquity form a single global system tuned to the size of the Earth and Moon. They are linked by fractions, both internally as divisions and multiples of their measures (*see table below*), and externally between modules (*see table by John Neal lower, opposite*). For example Assyrian and Greek measures are in the ratio 9:10, while Roman and Greek compare as 24:25, so that Assyrian and Roman are 15:16. The Belgic foot is 6:5 of the Assyrian and 9:8 of the Roman. The names of the various modules are often misleading as systems were regularly used simultaneously. Note the use of harmonic intervals.

It was Newton who reignited modern interest in ancient metrology. Seeking an accurate value for the size of the Earth he turned to the tradition that the Jewish 'sacred' cubit was a fraction of the Earth's radius (it is the 10 millionth part). In the years that followed, meticulous surveys of Egyptian and Greek sites by John Greaves, Francis Penrose, and William Flinders Petrie revealed small but regular variations in the individual modules. Petrie and, later, Livio Stechini noted that measures regularly varied around the 170th and 450th parts. John Michell and John Neal later refined and explained these.

The first variation in values results from the difference between the two values of π in the ancient world, 25/8 and 22/7, which works out as 175/176. Vitruvius describes a Roman odometer as having wheels four feet in diameter which mark a length of 12.5 feet on each revolution. Thus, although measured in the shorter Roman foot, a truer distance would have been found in the same number of longer Roman feet, longer by 176/175. The fraction shows up elsewhere, for instance the *shaku*, or traditional Japanese foot, which to this day relates to the English as 175:176.

The second variation may be derived from the non-spherical shape of the Earth. The ancients were aware that the equatorial radius exceeds the polar radius, meaning that degrees of latitude over the surface shorten towards the poles. The ratio between polar:mean:equatorial values was modelled in the ancient survey as 880:882:883, explaining the often noted 440:441 variation in particular units (sometimes termed the 'long' and 'short' foot). The mean radius occurs around 51°, known as the meridian degree.

These conversions kept linear measures harmonic to angular ones. For instance one second of arc at the Earth's surface at the meridian degree is 100 Greek feet. The English (and US) mile is likewise geodetic (*see page 327*).

		Digit	Inch	Palm	Hand	Span	Foot	Cubit	Step	Yard	Pace	Fathom
Digit	=	1	3/4	1/4	3/16	1/12	1/16	1/24	1/40	1/48	1/80	1/96
Inch	=	4/3	1	1/3	1/4	1/9	1/12	1/18	1/30	1/36	1/60	1/72
Palm	=	4	3	1	3/4	1/3	1/4	1/6	1/10	1/12	1/20	1/24
Hand	=	16/3	4	4/3	1	4/9	1/3	2/9	2/15	1/9	1/15	1/18
Span	=	12	9	3	9/4	1	3/4	1/2	3/10	1/4	3/20	1/8
Foot	=	16	12	4	3	4/3	1	2/3	2/5	1/3	1/5	1/6
Cubit	=	24	18	6	9/2	2	3/2	1	3/5	1/2	3/10	1/4
Step	=	40	30	10	15/2	10/3	5/2	5/3	1	5/6	1/2	5/12
Yard	=	48	36	12	9	4	3	2	6/5	1	3/5	1/2
Pace	=	80	60	20	15	20/3	5	10/3	2	5/3	1	5/6
Fathom	=	96	72	24	18	8	6	4	12/5	2	6/5	1

Dances of the Planets

MERCURY - VENUS

MERCURY - EARTH

MERCURY - MARS

MERCURY - CERES

MERCURY - JUPITER

MERCURY - SATURN

VENUS - EARTH

VENUS - MARS

VENUS - CERES

VENUS - JUPITER

VENUS - SATURN

EARTH - MARS

402

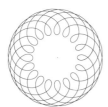

EARTH - CERES

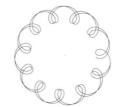

EARTH - JUPITER

EARTH - SATURN

EARTH - URANUS

MARS - CERES

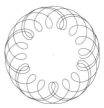

MARS - JUPITER

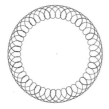

MARS - SATURN

MARS - CHIRON

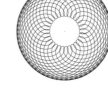

CERES - JUPITER

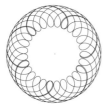

CERES - SATURN

CERES - CHIRON

JUPITER - SATURN

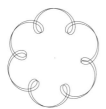

JUPITER - URANUS

JUPITER - NEPTUNE

JUPITER - PLUTO

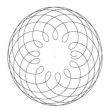

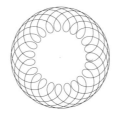

SATURN - URANUS

SATURN - NEPTUNE

SATURN- PLUTO

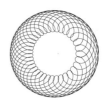

CHIRON - URANUS

CHIRON - NEPTUNE

CHIRON - PLUTO

URANUS - NEPTUNE

URANUS - PLUTO

NEPTUNE - PLUTO

INDEX

Bibliography & Further Reading

Deryck Cooke, *The Language of Music*, Oxford 1959
Harold Coxeter, *Regular Polytopes*, Dover 1974
Keith Critchlow, *Order in Space*, Thames & Hudson 1969
   Time Stands Still, Floris Books 2007
   Islamic Patterns, Thames & Hudson 1983
Peter Cromwell, *Polyhedra*, Cambridge 1999
Michael J. Crowe, *Theories of the World from Antiquity to the*
   Copernican Revolution, Dover 1990
Alain Danielou, *Music and the Power of Sound*, Inner Traditions 1995
David Fideler, *Jesus Christ, Sun of God*, Quest Books 1996
William Fleming, *Art, Music and Ideas*, Holt 1970
Jocelyn Godwin, *Harmonies of Heaven & Earth*, Thames & Hudson 1995
   Cosmic Music, Inner Traditions 1989
Joseph Goold, C. E. Benham, R. Kerr and L. R. Wilberforce, *Harmonic*
   Vibrations, Newton & Co. 1909
Kenneth Sylvan Guthrie & David Fideler, *The Pythagorean Sourcebook*,
   Phanes Press 1987
George Hart, (everything polyhedral) www.georgehart.com
Richard Heath, *The Matrix of Creation*, Inner Traditions 2004
   Sacred Number, Inner Traditions 2007
Robin Heath, *Sun Moon and Earth*, Wooden Books / Walker & Co. 2006
Herman Helmhotlz, *On the Sensations of Tone as a Physiological Basis*
   for the Theory of Music, Dover 1954
Alan Holden, *Shapes, Space and Symmetry*, Dover 1992
Georges Ifrah, *The Universal History of Numbers*, Harvill Press 1998
Jamie James, *The Music of the Spheres*, Grove Press / Little, Brown 1993
James Jeans, *Science and Music*, University Press 1937

Hans Jenny, *Cymatics*, Macromedia Press 2001
Robert Kanigel, *The Man Who Knew Infinity*, Abacus 1992
Hazrat Inayat Khan, *The Mysticism of Sound and Music*, Shambhala 1996
Arthur Koestler, *The Sleepwalkers*, Penguin 1989
Robert Lawlor, *Sacred Geometry*, Thames & Hudson 1982
Ernest McClain, *The Myth of Invariance*, Shambhala 1978
John Michell, *The New View Over Atlantis*, Thames & Hudson 1986
   The Dimensions of Paradise, Inner Traditions 2008
   How the World is Made, with Allan Brown, Thames & Hudson 2009
Angela Moore, *In Love with Venus*, Squeeze Press 2009
Guy Murchie, *Music of the Spheres*, Dover Books, 1961
John Neal, *All Done with Mirrors,* Secret Academy 2000
Scott Olsen, *The Golden Section*, Wooden Books / Walker & Co. 2006
Guy Lyon Playfair & Scott Hill, *Cycles of Heaven*, Souvenir Press 1978
Martin Rees, *Just Six Numbers*, Phoenix 2000
Michael Schneider, *A Beginner's Guide to Constructing the Universe*,
   Harper Perennial / Avon 1995
Arnold Schoenberg, *Fundamentals of Music Composition*, Faber, 1973
Joachim Schultz, *Movements and Rhythms of the Stars*, Floris Books 1986
Malcolm Stewart, *Patterns of Eternity*, Floris Books 2009
William Stirling, *The Canon*, Elkin Matthews 1897
Gordon Strachan, *Jesus the Master Builder*, Floris Books 2000
Daud Sutton, *Ruler and Compass*, Wooden Books / Walker & Co. 2009
David Wade, *Symmetry*, Wooden Books / Walker & Co., 2006
Magnus Wenninger, *Polyhedron Models*, Cambridge 1974
   Spherical Models, Cambridge 1979 & *Dual Models,* Cambridge 2003
and ... www.woodenbooks.com